The Stargazer's Guide to the Universe

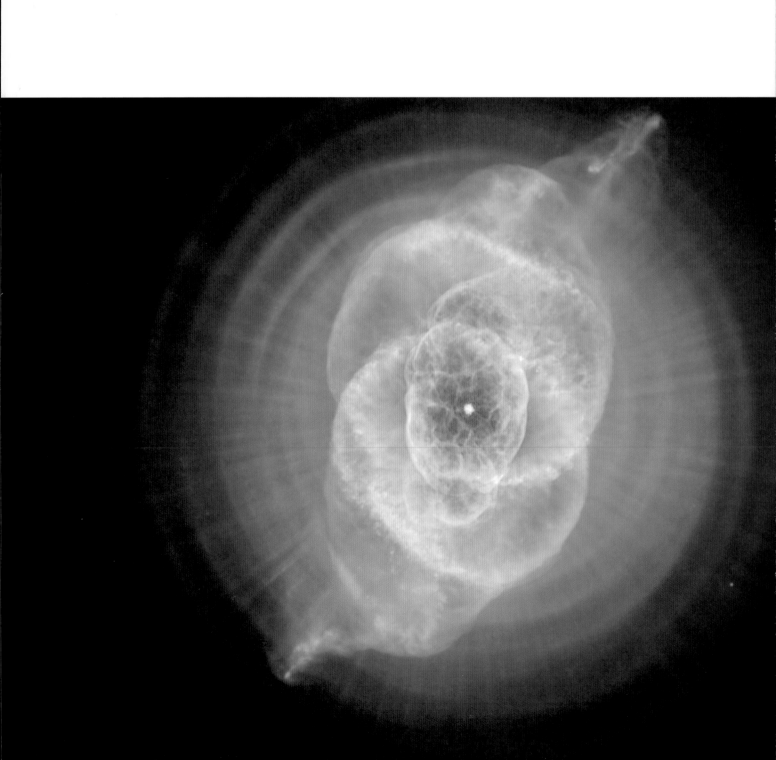

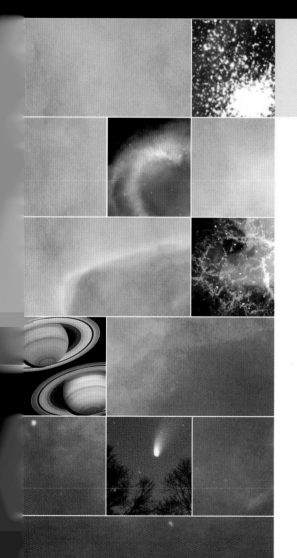

The Stargazer's Guide to the Universe

A complete visual guide to interpreting the cosmos

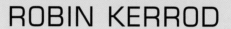

ROBIN KERROD

BARRON'S

A QUARTO BOOK

First edition for North America published in 2005
by Barron's Educational Series, Inc.

All inquiries should be addressed to:
Barron's Educational Series, Inc.
250 Wireless Boulevard
Hauppauge, New York 11788
www.barronseduc.com

ISBN-13: 978-0-7641-5844-5
ISBN-10: 0-7641-5844-9
Library of Congress Catalog Card No. 2004111725

QUAR.ITU

Conceived, designed, and produced by
Quarto Publishing plc
The Old Brewery
6 Blundell Street
London N7 9BH

Project Editor Susie May
Art Editor Claire Van Rhyn
Assistant Art Director Penny Cobb
Designer Rachel O'Dowd
Illustrator Kuo Kang Chen
Map makers Costronics Electronics
Copy Editor Victoria Leitch
Indexer Richard Emerson
Proofreader Robert Harries
Art Director Moira Clinch
Publisher Paul Carslake

Reproduction by Modern Age Repro House Ltd, Hong Kong
Printed in China by SNP Leefung Printers Limited

9 8 7 6 5 4 3 2 1

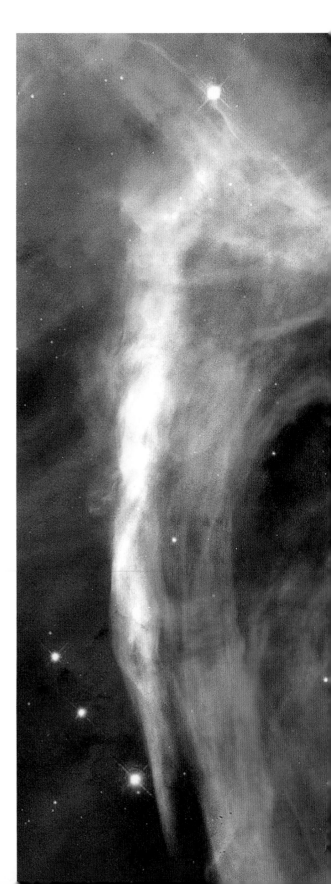

Contents

Introduction

Every clear night of the year, we are treated to one of nature's greatest spectacles—the star-studded night sky. Myriad stars shine down on us out of the deep, dark, velvety blackness of space.

People have been gazing at the night sky—the heavens—in awe and wonder since the beginning of civilization, more than 5,000 years ago. When they began studying the night sky in earnest and recording the changes that occur, they laid the foundations of astronomy, the now meticulous scientific study of the heavens and the heavenly bodies.

We all become astronomers of a sort, when we raise our eyes to the heavens and familiarize ourselves with the comings and goings that occur there as the months go by. We see the patterns of stars—the constellations—wheeling across the sky during the night, appearing and disappearing with the seasons.

Certain bright stars can be seen wandering among the otherwise fixed stars of the constellations. But these wandering stars are not stars at all, but planets, very much closer bodies that circle in space around our local star, the Sun. They form part of the Sun's family, or Solar System. Earth, too, is a planet, circling around the Sun once a year. We know now that planets circle around other stars, too.

All the bodies we see in the night sky—stars and planets, Earth, as well as other bodies such as the Moon and comets—belong to the universe. The universe means all that exists, or has existed, or will ever exist.

The night sky provides us with a window on the universe. Using powerful telescopes on the ground and also in space, professional astronomers look through this window into the

depths of space and discover a celestial menagerie of bizarre bodies such as quasars and blazars and incredible phenomena such as gravitational lensing. They are at last beginning to fathom what the universe is really like and what makes it tick.

Astronomers have discovered that all the stars we see in the sky belong to a great star island in space, which we call the Galaxy. There are many such star-island galaxies in space. On the broadest scale, scattered groups of galaxies—and the space between them—make up the universe.

You don't have to be a professional astronomer to appreciate the most tangible results of their work—the pictures, or images they acquire through their telescopes. Whether acquired by photography or, more usually these days, electronically, these images reveal how extraordinarily beautiful our universe is. Indeed, astronomical imagery has become an art form in its own right.

The Stargazer's Guide to the Universe focuses on some of the most stunning images produced to date. They include fabulous images from the Hubble Space Telescope and from giant ground-based telescopes such as the VLT (Very Large Telescope) in Chile, which is so powerful that it could probably detect an astronaut walking on the Moon!

But this book does more than show pretty astronomical pictures. It peers beyond the visual beauty to reveal the astronomical highlights in the images. Each image is chosen in sequence as part of a story, which explains the nature of the stars, the galaxies, and the Solar System. Words and images combine to present a stunning and digestible guide to our universe.

How to Use This Book

The primary purpose of *The Stargazer's Guide to the Universe* is to present some of the most outstanding images we have of the universe, and to analyze them astronomically.

CHAPTER 1 Introducing the Universe
sets the scene. It provides a general overview of the universe, considering how it came into being, how it is made up, and how it works.

CHAPTER 2 Exploring the Universe takes a detailed look at how telescopes work and at the way they acquire images. It looks in particular at the new generation of telescopes at ground observatories and at the space observatories and probes that carry instruments into orbit around Earth and beyond.

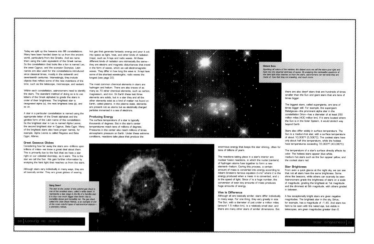

Introducing the Topic

A brief, 6-page information section precedes each of chapters 3, 4, and 5. It provides essential background information about the subject covered (stars, galaxies, the Solar System) in the chapter.

CHAPTER 3 Looking at Stars focuses on those pinpricks of light we see in the night sky that are actually great globes of incandescent gas like our own local star, the Sun. We see them as pinpricks only because they are so very far away.

CHAPTER 4 Looking at Galaxies zeroes in on the great star islands, or galaxies, that make up the universe. We can see only three galaxies in the night sky with the naked eye. All the others are too remote to be seen, some so far away that their light takes billions of years to reach us.

CHAPTER 5 Looking at the Solar System looks much nearer home, focusing on the Sun and the collection of heavenly bodies that circle in orbit around it. And what an extraordinary collection it is! It includes giant-sized gaseous planets such as Jupiter, dwarf rocky planets such as Earth, moons, and tiny icy chunks that reveal themselves as comets when they approach the Sun.

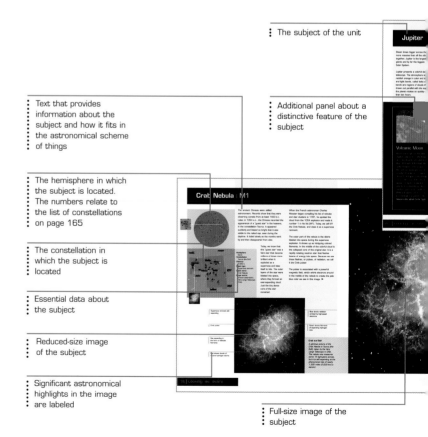

The subject of the unit

Additional panel about a distinctive feature of the subject

Text that provides information about the subject and how it fits in the astronomical scheme of things

The hemisphere in which the subject is located. The numbers relate to the list of constellations on page 165

The constellation in which the subject is located

Essential data about the subject

Reduced-size image of the subject

Significant astronomical highlights in the image are labeled

Full-size image of the subject

Text that provides information about the history and methods of mapping the skies

One of the pair of hemisphere maps, with the constellations

Diagram of the celestial sphere, used to pinpoint the postion of stars in the sky

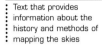

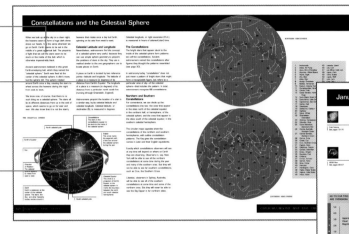

Reference Maps

A brief, 4-page information section precedes the monthly maps, providing information about mapping methods. The monthly maps show which heavenly bodies are visible during each month, and refer the reader to the pages where some of them are featured in Chapters 3 and 4.

Map of the constellations visible near the meridian during the featured month

Miniature image of the subject indicated and reference to the page where it is featured in the book

Key to star types and phenomena indicated on the map

REFERENCE SECTION Mapping the Skies presents the changing aspects of the heavens we see every month throughout the year. This section shows how the constellations dovetail together, reinforcing the constellation location maps in Chapters 3 and 4.

Further useful information is provided by the **Glossary** of astronomical terms used in the book and a list of major **Milestones in Astronomy**, from the time the priest–astronomers of the Middle East first turned their eyes skyward in an attempt to comprehend the mysteries of the heavens.

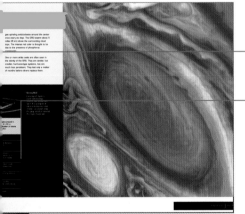

Essential data about the subject

Diagram showing the orbit of the subject

Analyzing the Image

The salient features of a typical 2-page unit in the core, analytical sections of the book, presenting an image of the subject, a labeled reduced version of the image, and keys to the constellation and celestial hemisphere in which the subject is located.

Chapter 1

Introducing the Universe

We know today that our universe is vast—far bigger than we can ever imagine or comprehend. It appears to go on for ever and ever. But this concept of a boundless universe is a relatively recent one, dating from the early years of the last century.

In classical Greece, philosophers such as Aristotle (300s B.C.) reasoned that a perfectly spherical Earth had to be the center of the universe. Later, in about 150 B.C., an astronomer in Alexandria named Ptolemy summed up this ancient Earth-centered concept, which became known as the Ptolemaic Universe.

In the Ptolemaic Universe, Earth was at the center. All the other heavenly bodies revolved around it—the Sun, the Moon, and the five planets then known (Mercury, Venus, Mars, Jupiter, and Saturn). These bodies all circled Earth at different distances and different times. Encompassing them all was a great celestial sphere that carried the background stars. This sphere was the universe.

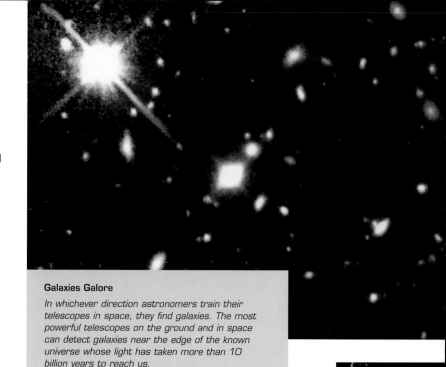

The concept of the Ptolemaic Universe held sway until the time of the Renaissance in Europe—the rebirth of learning, the arts, and the sciences that took hold in the fourteenth and fifteenth centuries. But it was not one of the leading astronomers of the day who ushered in an astronomical revolution but a Polish cleric with a penchant for astronomy.

Named Nicolaus Copernicus, he put forward the idea that the motions of the heavenly bodies could better be explained if the Sun and not Earth were at the center of things. He set down his ideas in the book *Concerning the Revolution of Celestial Spheres*. It was produced as he lay on his deathbed in 1453.

Copernicus's idea of a Sun-centered (Solar) System caused an indignant reaction, especially from the Church, which considered it heretical. It took more than a century for the Copernican Universe to be widely accepted. Many people suffered because of the Church's attitude, including the Italian genius of a scientist Galileo Galilei.

Galaxies Galore

In whichever direction astronomers train their telescopes in space, they find galaxies. The most powerful telescopes on the ground and in space can detect galaxies near the edge of the known universe whose light has taken more than 10 billion years to reach us.

The Revolution Continues

It was Galileo who continued the astronomical revolution when he built a telescope and trained it on the heavens during the winter of 1609–10. And what sights he saw—the phases of Venus, moons of Jupiter, and a Milky Way teeming with stars in their millions.

As bigger and better telescopes were built, it became apparent that the universe was much greater in size than anyone had imagined. The discovery of a new planet (Uranus) beyond Saturn, by English musician-turned-astronomer William Herschel, in 1781, effectively doubled the size of the known Solar System, and it nearly doubled in size again when another new planet (Neptune) was discovered in 1846.

A few years earlier, astronomers had begun, for the first time, to gain an idea of just how far away the stars are. In 1838, German astronomer Friedrich Bessel succeeded in measuring the distance to a star named 61 Cygni, in the constellation Cygnus. He used a principle known as parallax, in which nearby objects seem to move against a distant background when viewed from different angles. He worked out that 61 Cygni lies an incredible 65 million million miles (105 million million km) away, a figure that astonished him

and his fellow astronomers. They began to appreciate for the first time just how enormous the universe must be.

Beyond the Galaxy

By now, astronomers had also realized that all the stars lie at different distances from us, and they began to visualize the universe as a bulging, convex lens, which they called the Galaxy. Not until U.S. astronomer Edwin Hubble began investigating outer space, with the giant 100-inch (2.5-m) Hooker telescope at Mount Wilson Observatory, in California, did it become evident that there were other galaxies in space far beyond our own.

This takes us, broadly speaking, to where we are today. Stars group together to form galaxies, and galaxies group together to form the universe.

The Fleeing Galaxies

Another fundamental fact Hubble and his fellow astronomers found early in the twentieth century was that the galaxies were not only very far away, but were also rushing away from us—and one another—at incredible speeds. And the farther they were away, the faster they appeared to be traveling. This happened in all directions in space, leading to the idea that the universe is getting bigger. This concept is called the expanding universe.

If the universe is expanding, it follows that it was smaller in the past. By measuring the rate at which it

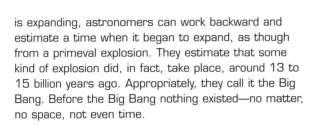

is expanding, astronomers can work backward and estimate a time when it began to expand, as though from a primeval explosion. They estimate that some kind of explosion did, in fact, take place, around 13 to 15 billion years ago. Appropriately, they call it the Big Bang. Before the Big Bang nothing existed—no matter, no space, not even time.

After the Big Bang

Amazingly, cosmologists—astronomers who study the origin and evolution of the universe—believe that they can trace what has happened to the universe since the Big Bang, except for the first 10^{-43} seconds. It was only after this time that the laws of physics as we know them today came into play.

At first, the embryonic universe was superhot and filled with radiation. Then matter began to form, and fundamental forces such as gravity came into play. But it took some 300,000 years for atoms to form— atoms are the building blocks from which all matter is made up.

The first atoms formed were the light gases hydrogen and helium, and they still dominate the universe today. They are the stuff that stars are made of, and they clumped together to form the first stars when the universe was between one and two billion years old.

Nature of the Universe

Simply stated, the universe consists mainly of empty space, with pockets of matter (stars, galaxies, planets) scattered here and there. The behavior of matter and the universe in general is governed by a set of fundamental forces.

Matter—in the stars, the planets, everyday objects, our bodies—can take three main physical states. It can be solid (like rock), liquid (like water), or gas (like air). These are called the three states of matter.

Whatever its state, matter is made up of tiny particles called atoms. The different chemical elements, such as hydrogen, copper, and iron, are made up of different kinds of atoms. There are about 90 different chemical elements in nature. Most forms of matter, however, exist as chemical compounds, combinations of two or more elements. Water is a combination of hydrogen and oxygen; table salt, a combination of sodium and chlorine.

Stars in the Galaxy

Billions of stars of different size, mass, and brightness populate our home Galaxy. This beautiful starscape is to be found in the constellation Carina in far southern skies. The stars in the central cluster are much more massive than our own Sun and very much younger—some are only about 300,000 years old (compared with the Sun's age of nearly 5 billion years).

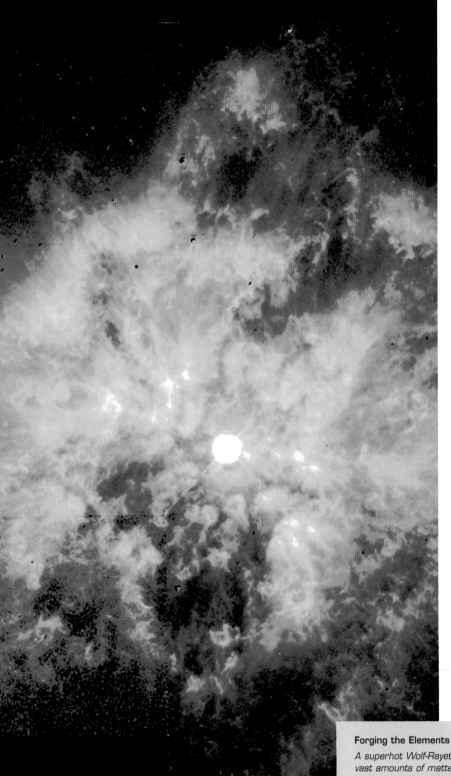

Atoms are the smallest particles of an element that can exist by themselves. But they are not the smallest particles. They are themselves made up of still smaller particles, such as protons, neutrons, and electrons. Most atoms consist of a nucleus (center) containing protons and neutrons, with a cloud of electrons circling around it.

Fundamental Forces

Four basic, or fundamental forces, are in the universe and govern its behavior. Two of them—the strong force and the weak force—act only at very close quarters within the nucleus of atoms to help keep its particles in place. The electromagnetic force holds atoms together. It provides the force of attraction between the nucleus, which has a positive electric charge, and the circling electrons, which have a negative electric charge.

The electromagnetic force also acts within our familiar everyday world, being responsible for the phenomena of electricity and magnetism.

A Matter of Gravity

Of all the fundamental forces, the weakest is gravity, which is the force of attraction one lump of matter has for another. But, unlike the other fundamental forces, gravity acts over vast distances and is responsible for holding the universe together. The more massive a body is, the more powerful is its gravitational attraction.

Earth's gravity is what keeps our feet on the ground and holds the Moon in its orbit around our planet. The Sun's much more powerful gravity holds the planets in their orbits, acting over distances of billions of miles.

Forging the Elements

A superhot Wolf-Rayet star, ten times hotter than the Sun, ejects vast amounts of matter into space during its spectacularly brilliant but short life. It is in the heart of such stars that all the heavy elements in the universe are forged out of the light elements hydrogen and helium that were created shortly after the Big Bang. When the stars eject matter into space, they scatter the heavy elements far and wide.

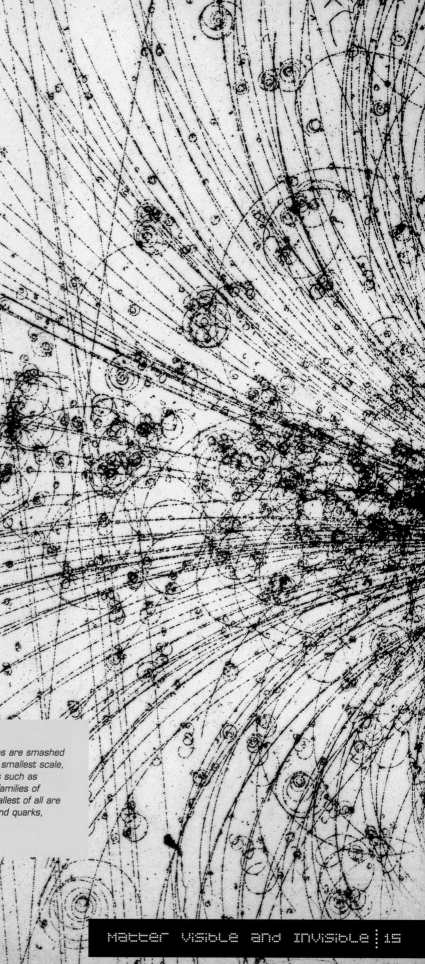

The gravity of our Galaxy keeps the Sun and the other stars circling around its center.

The mutual gravitational attraction between our Galaxy and others nearby holds them together in a loose group. In turn, groups of galaxies collect together in clusters, and clusters gather into superclusters. On the largest scale, gravity brings together the superclusters to form the universe.

Fate of the Universe

We think we know how the universe began—in a Big Bang—and we know what it is like now. But what will happen to it in the future? Will it end, and if so how? Or will it go on and on for ever and ever?

The key factor in deciding which course the universe will ultimately take is gravity—the combined gravity of all the matter in the universe. If there is enough matter and its combined gravity is great enough, it could eventually rein in the galaxies that are at present rushing away from one another. It could stop the universe expanding and in time pull back the galaxies and cause the universe to shrink. Ultimately, all the matter in the universe could come together in a reversal of the Big Bang—cosmologists refer to such an event as a Big Crunch. If, on the other hand, the combined gravity of all the matter in the universe is not great enough, then there will be nothing to stop the outrush of the galaxies, and the universe will go on expanding for ever.

Matter Visible and Invisible

The amount of matter in our universe, and its combined gravity, are crucial in deciding the universe's ultimate fate. When cosmologists estimate the amount

Particles Aplenty

Showers of particles are produced when atoms are smashed to pieces in particle accelerators. On the very smallest scale, the universe consists of particles—larger ones such as molecules, smaller ones such as atoms, and families of subatomic particles—smaller than atoms. Smallest of all are the fundamental particles such as electrons and quarks, which cannot be split up further.

of matter in the universe that we can see, as stars and galaxies, they find that there is nowhere near enough to stop the universe from expanding.

But there is compelling evidence that there may be a great deal of matter in the universe that we can't see. We call this invisible matter dark matter. Indeed, it is estimated that visible matter makes up only about 10 percent of the matter in the universe and dark matter 90 percent. Astronomers find evidence for the presence of dark matter in spiral galaxies, for example. The way they rotate and retain their structure over time suggests that they must contain more matter than that of their visible stars, which has to be dark matter. Likewise, the way galaxies remain in clusters despite their high speed shows that invisible, dark matter must be present to exert the necessary gravitational constraint.

MACHOs and WIMPs

Because we can't see it or detect it directly, we don't know what dark matter is like. Astronomers have postulated that it probably takes different forms. In the halos, or outer regions of galaxies, it could take the form, for example, of dead stars or black holes. These invisible objects are called MACHOs, or massive compact halo objects.

Dark matter might also take the form of elusive atomic particles with a tiny mass that hardly interact at all with ordinary matter. They are called WIMPs, or weakly interacting massive particles. Particles known as neutrinos, which we know do exist and that flood the universe, could fit the bill if they are found to have a slight mass, and there is some evidence to suggest that they have.

Scale of the Universe

We know that the universe is vast—bigger than we can ever imagine. But how big is that? Astronomers estimate that the farthest bodies they can detect—

Missing Matter

This distant cluster of galaxies, named Abell 1689, has a strong gravity that affects surrounding objects and even bends light rays. But there is not enough visible matter to account for the strength of its gravity. So astronomers figure that it must therefore contain huge amounts of dark matter.

remote embryonic galaxies—lie farther than 60 billion trillion miles away (about 100 billion trillion km).

It is quite impossible for anyone to comprehend distances such as 60 billion trillion, or 60,000,000,000,000,000,000,000 miles. Astronomical distances are so great that ordinary units of distance such as the mile and the kilometer are just too small.

They are too small to measure the distance to even the nearest star, Proxima Centauri, which lies more than 25 trillion miles (40 trillion km) away. It lies so far away that its light, traveling at a speed of 186,000 miles (300,000 km) per second, takes over four years to reach us.

This gives us a means of simplifying how we express distances in space. We use the distance light travels in a year as a unit of measurement, calling it the light-year. Using this unit, we can say that Proxima Centauri lies more than 4 light-years away—so much simpler than trillions of miles or kilometers. And those remote embryonic galaxies at the edge of the known universe? They lie more than 10 billion light-years away.

We use the light-year (equal to about 6 trillion miles or 10 trillion kilometers) as a unit to express distances

throughout this book. Most professional astronomers, however, use a unit called the parsec, which is equivalent to about 3.3 light-years.

A Grand Tour

To get an idea of the scale of the universe, imagine that we set off on a grand tour of space in an interstellar spaceship that can travel at the speed of light. Leaving Earth, we shoot past the Sun in about 8 minutes. Pluto flashes by after 5½ hours. Then comes a journey of more than 4 years to reach Proxima Centauri and its companion stars.

Heading into the heart of our Galaxy, it takes 25,000 years to reach its center, and another 50,000 years to reach its outer edge. More than 160,000 years would go by before we came to our neigboring galaxy, the Large Magellanic Cloud, and 2.5 million years before we reached the Andromeda Galaxy (the most distant object we can see in the sky with the naked eye).

As we continue on our journey, millions and billions of years go by as we pass galaxy after galaxy. After more than 10 billion years, we come upon the most remote galaxies astronomers can detect in their telescopes. We will have reached the edge of the known universe.

Chapter 2

Exploring the Universe

Our early ancestor astronomers began exploring the mysterious universe of stars and space just with their eyes. Then, around four centuries ago, Galileo pioneered telescopic observation. Astronomical knowledge advanced by leaps and bounds as bigger and better telescopes were built. Today, astronomers use huge telescopes at ground observatories to gather the faint light from remote galaxies that has been traveling for billions of years to reach us.

Space telescopes are now playing an increasingly important role in exploring the universe. The Hubble Space Telescope in particular has revealed the beauty of the universe as never before. Other space telescopes are showing us what the universe looks like at invisible wavelengths, such as infrared. Nearer home, space probes traveling through interplanetary space have revolutionized our study of the planets and other bodies in the Solar System.

Above the Clouds

Telescopes are the main instruments astronomers use to explore the universe. These three telecope domes are located at one of the world's finest observatories—Kitt Peak National Observatory, near Tucson, in Arizona. The dome on the right houses the Mayall 157-inch (4-m) reflector. Kitt Peak is ideally located for an observatory. It is perched on a mountaintop 6,875 feet (2,095 m) high, where it is above the clouds. The climate is warm and dry, and there is little air pollution.

Visible Astronomy

Stars pour out energy into space in the form of electromagnetic radiation—minuscule electric and magnetic disturbances that ripple through space in the form of waves. Our eyes are sensitive to a range of waves that together make up visible light. In their work, astronomers study primarily the visible light from stars and galaxies, supplemented by studies of the invisible waves these bodies give out (see page 22). The prime instrument they use to study starlight is the telescope.

Lenses and Mirrors

The two main types of telescope are refractors and reflectors. Refractors use glass lenses to gather and focus incoming light; reflectors use mirrors.

At the world's leading observatories, astronomers use large reflectors for observing. Reflectors can be built in much bigger sizes than refractors because their mirrors can be supported from behind. The lenses of refractors can be supported only around the edge, which causes increasing distortion as the lens size increases. The world's biggest refractor (at Yerkes Observatory, Wisconsin) has a light-gathering lens only 40 inches (102 cm) across. Compare this with the Keck telescopes on Hawaii, which have light-gathering mirrors 33 feet (10 m) across.

Capturing the Image

In a reflector, incoming light is received by a curved primary mirror. This reflects light onto a secondary mirror above it, which in turn focuses the light into an eyepiece (for eye-viewing) or onto some kind of detector. In the Cassegrain-focus system favored in many telescopes, the secondary mirror reflects light back down the telescope tube and through a hole in the primary. The most popular amateur setup is the Newtonian, in which the eyepiece is located at the side of the body tube at a convenient height for viewing.

These days, professional astronomers seldom look through their telescopes. Instead, they use them as cameras and gather incoming starlight on photographic film or on microchip devices called CCDs. The CCDs are similar to those used in digital and video cameras. They are made up of minute pixels (picture elements), which acquire an electric charge when light strikes them. A pattern of charges builds up on the chip that represents the incoming pattern of light. A computer reads the electronic pattern and can convert it into a visible image.

On Film

This galaxy, NGC 2841, is found in Ursa Major. It is a spiral rather like our own Galaxy. Until quite recently, images of astronomical objects were captured on photographic film. They were photographed in black and white, as here. Simulated color pictures could be built up by combining images taken through different-colored filters.

The Northern Twin

Bird's-eye view of the Gemini North telescope at Mauna Kea Observatory in Hawaii. At the bottom is the 26-foot (8-m) primary mirror. At the top is the secondary mirror assembly. The skeletal framework is typical of modern telescopes, providing structural strength for the lightest weight. Gemini South in the Chilean Andes has an identical design.

The advantage of film and chips is that they can in effect store incoming light. If they are exposed for a long time, they will gradually gather faint starlight and produce brighter images. For long exposures, telescopes must be driven so that they can follow the stars as they wheel overhead.

The Big Eyes

Broadly speaking, the increase in astronomical knowledge relates directly to the size of the telescopes astronomers use. For example, in the 1920s, Edwin Hubble used the new 100-inch (254-cm) Hooker telescope at Mt. Wilson Observatory near Los Angeles to confirm the existence of galaxies beyond our own. In 1948, an instrument double that size was completed at nearby Mt. Palomar Observatory—the 200-inch (508-cm) Hale telescope—and remained at the cutting edge of astronomy for decades.

Though optically still outstanding, the Hale telescope is suffering increasingly from light pollution from the urban sprawl of Los Angeles. The finest observatories today are sited well away from urban areas, high up on mountaintops. Particularly outstanding are Mauna Kea Observatory on the Big Island of Hawaii and the Paranal Observatory in Chile in South America.

Mauna Kea is the site of the twin Keck telescopes, which have light-gathering mirrors no less than 33 feet (10 m) in diameter. Unusually, these mirrors are not made as a single unit. They are made up of 36 separate segments, each individually supported by computer-controlled rams so they can together maintain perfect curvature. Mauna Kea is also the site of the 27-foot (8.2-m) Gemini North telescope, which has a twin (Gemini South) on the mountain Cerro Pachon in the Andes.

The Paranal Observatory is located on a mountaintop in one of the driest places on Earth—the Atacama Desert in Chile. It is the home of what is currently the most powerful telescope setup in the world, the Very Large Telescope (VLT). It comprises four identical telescopes, with mirrors 27 feet (8.2 m) across. They can be used individually, in combination with one another, and in combination with other smaller telescopes on site. When all the instruments are combined, they have the equivalent light-gathering power of a mirror 660 feet (200 m) across.

Invisible Astronomy

Stars give off their energy not only as visible light rays, but also as rays that we can't see. These invisible rays are, like visible light, different kinds of electromagnetic radiation. They differ from one another in their wavelength. Wavelengths are always expressed in metric measurements as meters or fractions of meters. Some wavelengths are so short that they are measured in nanometers, or billionths of a meter.

Gamma rays have the shortest wavelengths—around 0.01 nanometers. Next come X-rays, then ultraviolet rays—rays with shorter wavelengths than the violet rays in visible light. The invisible rays with longer wavelengths than the red rays in visible light are infrared rays. Next come the longer-wavelength microwaves, and finally radio waves, which have the longest wavelengths of all—up to several miles.

The Radio Window

Studying the invisible rays stars give out from the ground presents a problem because most rays are absorbed as they travel through the atmosphere. Only radio waves make it down to the ground, through what is called the atmosphere's radio window. In 1931, Bell Telephone engineer Karl Jansky first discovered that radio waves reached Earth from space. In so doing, he laid the foundations of what has become one of the most exciting branches of astronomy—radio astronomy.

Astronomers study the radio waves from the heavens using radio telescopes. These instruments look nothing like ordinary optical telescopes. Most take the form of a huge metal dish with an antenna located above it. The dish gathers the radio signals from space and focuses them on the antenna. The signals are then fed to a control room, where they are amplified (strengthened) and analyzed by computer.

The radio signals that radio telescopes capture are, of course, invisible, so they cannot be directly imaged. But computers can translate and display the signals as images that are equivalent to the pictures we might see if our eyes were sensitive to radio waves. The pictures are displayed in false (not true-to-life) colors, of course. Color combinations are chosen so as to bring out the maximum detail in the image.

Big Dishes

Radio telescopes really are huge. The Arecibo radio telescope on the Caribbean island of Puerto Rico has a dish 1,000 feet (300 m) across, built into the top of a mountain. It relies on Earth's rotation to scan the sky. With a 330-foot (100-m) dish, the Effelsberg radio telescope in Germany is the biggest steerable dish in the world.

But much more powerful is the Very Large Array (VLA) near Socorro, New Mexico. The VLA is made up of 27 separate 82-foot (25-m) dishes, mounted on rails so that they can be moved into different configurations. Coupled together electronically, they can synthesize a dish nearly 20 miles (30 km) across.

Into Space

We can study the radio waves coming from the heavens at ground level, but not the other invisible

Big Dish

The Effelsberg radio telescope, the biggest steerable single-dish instrument in the world. Located near Bonn, in Germany, it began operation in 1972. The 330-feet (100-m) diameter dish of the telescope is made up of more than 2,350 metal panels.

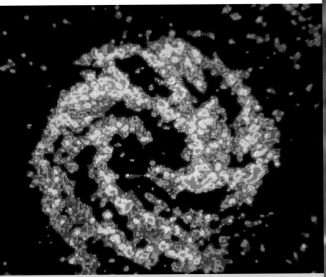

Radio Pictures

In the control room of a radio telescope, radio signals captured by the instrument are analyzed by computer and can be displayed as images. This picture shows a radio image of a head-on spiral galaxy. The false colors in the image reflect the radio brightness of different parts of the object.

rays, which the atmosphere absorbs. To gain access to these other rays, astronomers must launch instruments into space on satellites that orbit Earth. Launching telescopes and other instruments into space has the additional advantage that the universe can be viewed with greater clarity, because these instruments are above Earth's dirty and distorting atmosphere.

The invisible universe has now been studied in some detail at most invisible wavelengths. Studying gamma rays helps astronomers understand the most violent events in the universe, associated with pulsars, quasars, and black holes. X-ray studies reveal the hot spots of the universe, such as the remnants of supernovae, the gas spiraling into black holes, and the searing hot gas at the center of galaxy clusters.

Ultraviolet studies reveal the presence of young, hot stars, which glow brightest at such wavelengths. Infrared studies reveal the cooler regions of the universe, such as the failed stars we call brown dwarfs. Infrared wavelengths can also penetrate dust clouds to reveal what lies beyond.

But not all astronomical satellites study the universe at invisible wavelengths. The Hubble Space Telescope operates mainly in visible light (see page 24).

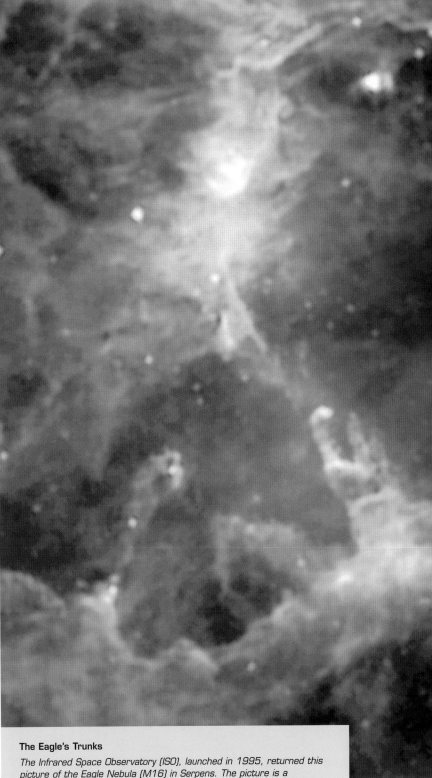

The Eagle's Trunks

The Infrared Space Observatory (ISO), launched in 1995, returned this picture of the Eagle Nebula (M16) in Serpens. The picture is a combination of two images taken at different infrared wavelengths and printed in blue and red. The Eagle Nebula was also the subject of one of the Hubble Space Telescope's most famous images, termed the "pillars of creation" (see page 58). These pillars—columns that astronomers call elephant trunks—are visible in this picture, just below center.

Hubble Space Telescope

More than any other astronomical instrument before or since, the Hubble Space Telescope (HST) has captured the imagination, not only of the scientific community, but of the public at large. The data and superlative images it has been returning since the 1990s have transformed our astronomical knowledge, as well as elevated space photography to a unique art form.

The HST is named after U.S. astronomer Edwin Hubble, who in the 1920s pioneered the study of the galaxies, proving conclusively that they lie far beyond our own star system, our own Galaxy. He also discovered that the galaxies are fleeing headlong away from us—and from one another—leading to the concept that the universe is expanding.

How the HST Works

The HST has caused an astronomical revolution, but optically its design is very conventional. It is a type of reflecting telescope known as a Cassegrain. Light gathered by its 95-inch (2.4-m) primary mirror is reflected up to a secondary mirror above. This mirror in turn reflects the light back the way it came and through a hole in the primary.

Behind the primary mirror, the light is focused on cameras and other instruments. Images are produced in the form of streams of electronic data, which are beamed back to the ground by radio antennae.

The HST operates mainly at visible light wavelengths, but it also has the capability of working at some ultraviolet and some infrared wavelengths as well.

Servicing the HST

The HST was designed so that it could be periodically serviced by astronaut–engineers from the space shuttle. Its instruments are contained in modules that can easily be replaced or refurbished.

Four servicing missions have taken place, in 1993, 1997, 1999, and 2002. The first was an emergency operation to correct the faulty optics of the original mirror assembly, which caused blurring of the images. No further shuttle-based servicing missions will be able to take place because of limitations on the use of the shuttle after the orbiter *Columbia* broke up on returning to Earth in February 2003.

During the servicing missions, all of the original instruments were replaced. The key instruments now in operation include the Wide-Field/Planetary Camera (WF/PC), Near-Infrared Camera and Multi-Object Camera (NICMOS), and the Advanced Camera for Surveys (ACS).

Producing the Images

The sensors used in the HST to record images are charge-coupled devices, or CCDs. These are also used extensively in ground-based telescopes (see page 20).

The beautiful color images of nebulae and galaxies taken by the HST do not come back from space as color pictures. They are combinations of separate

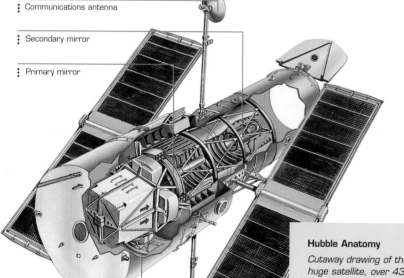

Communications antenna

Secondary mirror

Primary mirror

Wide field and planetary camera, Hubble's main camera

Solar array for power

Hubble Anatomy
Cutaway drawing of the Hubble Space Telescope (HST). It is a huge satellite, over 43 feet (13 m) long and 14 feet (4.3 m) in diameter. It has a mass of more than 12 tons (11 tonnes). It operates in a near-circular orbit about 350 miles (560 km) above Earth. Built at a cost approaching U.S.$1.5 billion, the HST was launched into space on April 24, 1990.

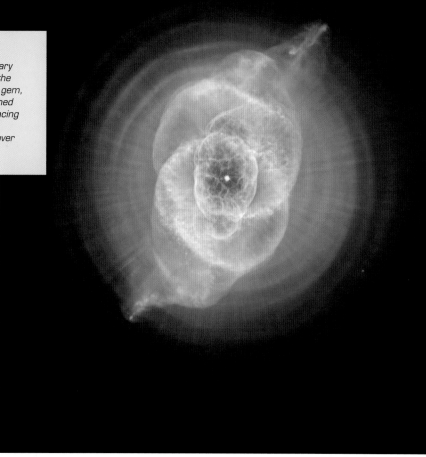

Into the Cat's-Eye

In 1995, the Hubble Space Telescope turned its Wide-Field/Planetary Camera on one of the most beautiful planetary nebulae we know—the Cat's-Eye in Draco (see page 65). In 2004, it revisited this celestial gem, using its supersensitive Advanced Camera for Surveys, which returned this extraordinary image. It shows a remarkable structure of interlacing rings, which are actually the way we see the spherical bubbles of glowing gas that the white dwarf star in the center has puffed off over thousands of years.

images taken through color filters, usually red, green, and blue. When combined together, the red, green, and blue light coming from an object synthesize a natural color image. This is similar to what happens on a TV screen—a natural color image is built up of a combination or red, green, and blue dots.

Hubble scientists also produce images in false color. They assign different colors to images taken through different filters so as to highlight different features.

Future Plans

Originally, it was planned that the HST would remain in operation until around the year 2010. But, with servicing missions on hold, the telescope will probably not last that long. It is not so much the instruments that will deteriorate; it is the positioning systems that allow it to lock onto specific targets. They depend primarily on sets of gyroscopes, which tend to fail after a time.

However, the next-generation space telescope is now being planned, which could be launched into space as early as 2011. It has been named the James Webb Space Telescope (JWST) for a former NASA administrator.

The JWST will be quite a different instrument from the HST. It will have a light-gathering mirror at least 20 feet (6 m) across, and it will operate mainly in the infrared. It will not be placed in Earth orbit, but in a gravitationally stable position in space some 930,000 miles (1,500,000 km) away. It will specifically target young stars, extrasolar planetary systems, and remote galaxies born in the early universe.

Colliding Galaxies

The Hubble Space Telescope excels in its study of galaxies, being able to see deeper into space—or farther back in time—than ground-based instruments. Here it has imaged a dense cluster of galaxies 8 billion light-years away. Studying the galaxies has revealed that perhaps as many as one in eight of them are colliding or have collided. This confirms that collisions between galaxies were much more common in the past than they are today.

Probing the Solar System

Icy Moon

*The Galileo probe returned this image
as it flew past Jupiter's moon, Europa.
The image shows (in false color) cracks
and ridges on Europa's ice-covered
surface. The brown streaks show
where material from underneath has
seeped through the cracks and stained
the ice. Europa, diameter 1,945 miles
(3,130 km), is one of the four large
Galilean moons that circle around
Jupiter.*

Just as orbiting space observatories have transformed our knowledge of stellar astronomy and the universe as a whole, so space probes have transformed our knowledge of objects much nearer home—in the Solar System.

Space probes are spacecraft that travel through the depths of the Solar System to explore planets, moons, and other bodies from close quarters. In 1959, the first probes (*Luna*) were sent to our neighbor in space, the Moon, as would be expected. Three years later, *Mariner 2* targeted Venus with some success. But it was *Mariner 4* in 1965 that showed the way ahead, taking close-up pictures of the Red Planet, Mars.

Since then, all the planets in the Solar System, except Pluto, have been explored from close quarters, with Mars being the most-favored destination. The moons of the planets have been targeted too, as have some comets and asteroids.

Escaping Earth

Broadly speaking, it is relatively simple to launch a satellite into orbit around Earth. We launch it on a rocket in the right direction and boost it to a speed of around 17,500 mph (28,000 km/h). And at this orbital velocity, it will go into orbit—in space but still bound to Earth by gravity.

Launching a probe into the depths of the Solar System is many magnitudes more difficult. First, it has to be launched from Earth at a speed of at least 25,000 mph (40,000 km/h). Only at this speed (Earth's escape velocity) can it break free from the clutches of gravity.

Next, the probe must be launched away from Earth in a very precise direction. Every body in the Solar System is moving, of course. So the probe must be aimed so that it will arrive at the same point in space as its target. This encounter may be many months or even years ahead. In working out the precise trajectory (path) the probe will need to take, account must be taken, for example, of the gravitational influences of other planets the probe might pass on the way.

Sometimes, the gravity of one planet is exploited deliberately to help a probe on its way to another. The probe is sent to fly close to the planet, and the planet's gravity makes the probe accelerate and loop into another trajectory that will take it on to its next target. Using this gravity-assist maneuver can greatly reduce journey times.

Probing the Inner Solar System

As yet, only one probe has been sent to Mercury—*Mariner 10* (1974). It revealed that the planet has a heavily cratered, almost Moonlike surface. *Mariner's* was a fly-by mission, in which its cameras photographed the planet as it flew past. In 2011,

Spirit Panorama

A magnificent panorama of the Martian surface, taken by the Mars exploration rover *Spirit*. It was assembled from 78 separate images taken by the probe's panoramic camera over a two-day period in March 2004, two months after *Spirit* began exploring the surface. *Spirit* sits in Bonneville Crater. The twin tracks it has made in the thin Martian soil show up clearly. Hardly visible in the middle distance where the tracks begin is its landing craft, named the Columbia Memorial Station to honor the crew of the shuttle orbiter which had perished the previous year.

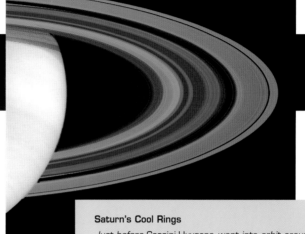

the *Messenger* probe is scheduled to go into orbit around the planet and photograph the whole surface.

We have gained much of our knowledge about Venus from *Magellan*, which began scanning the planet from orbit in 1990. It used radar to see through the perpetual clouds that otherwise hide the surface, which has been shaped by volcanoes. Venus proves to be quite a planet, with a body-crushing atmosphere, sulfuric acid clouds, and temperatures reaching 900°F (480°C).

Mars has been targeted by more probes than any other planet. In some respects it is similar to Earth, and is the only planet humans could conceivably live on. It is possible that some kind of life once existed there, when its climate was less harsh than it is now. The *Viking* landing probes in 1976 actually searched for traces of life, but in vain.

The year 2004 was an exceptional year for Mars exploration. *Mars Express* began returning the best images ever of the surface from orbit. The Mars rovers *Spirit* and *Opportunity* roamed across the surface, analyzing the rocks and soil and returning amazing images.

Probing the Outer Solar System

Study of the outer planets began with fly-bys of Jupiter by *Pioneer 10* in 1973 and *Pioneer 11* a year later. But it was the *Voyager 1* and *2* probes that revolutionized our knowledge of all four giant planets. Both probes explored Jupiter and Saturn (1979–1981), and then *Voyager 2* went on to target Uranus (1986) and Neptune (1989).

Saturn's Cool Rings
Just before Cassini-Huygens *went into orbit around Saturn in July 2004, its infrared spectrometer accurately measured the temperature of the planet's main rings. Here, the data has been depicted in a false-color image. Red regions represent a temperature of around -261°F (-163°C); green, around -297°F (-183°C); and blue, around -333°F (-203°C). In other words, the more transparent parts of the rings—the inner C ring and the Cassini Division—are warmest, and the more opaque sections—the A and B rings—are coolest.*

Galileo went into orbit around Jupiter in 1995, and *Cassini-Huygens* around Saturn in 2004. *Huygens* was a landing probe, which parachuted through the atmosphere of Saturn's large moon Titan early in 2005.

Space probes have also targeted some of the smallest bodies in the Solar System. On its way to Jupiter, *Galileo* returned the first pictures of asteroids. The asteroid Eros was the destination of *Near–Shoemaker*, which eventually landed on it in 2001.

Comets, too, have been explored. A flotilla of five probes encountered Halley's Comet in 1986, *Giotto* taking spectacular close-up pictures of its icy nucleus. *Stardust* is returning from an encounter with Comet Wild 2 early in 2004. Carrying samples of comet dust, it should return to Earth early in 2006, when the samples should be recovered. They could tell astronomers much about conditions in the early Solar System.

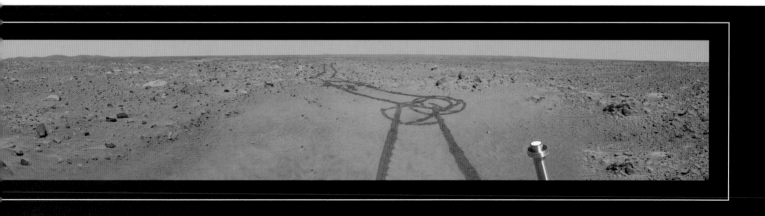

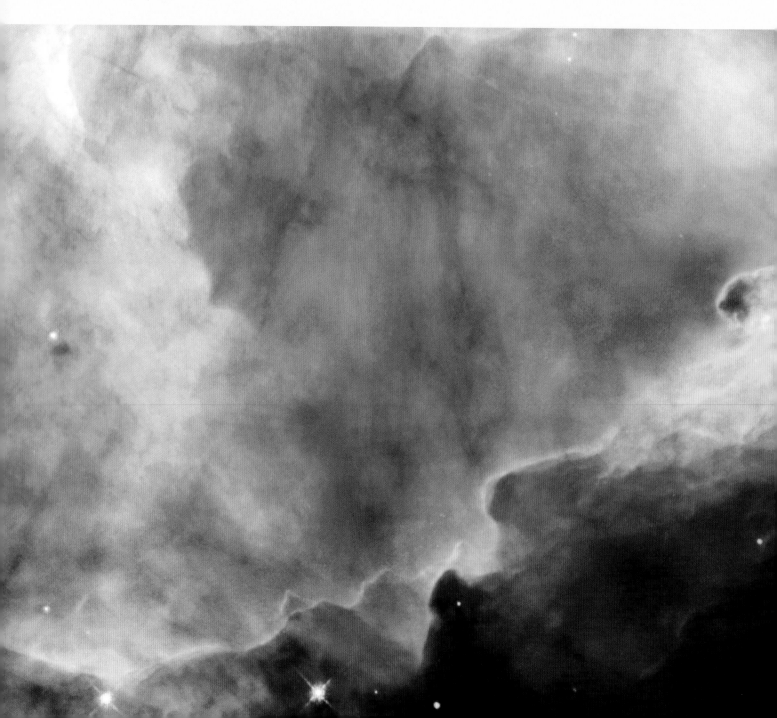

Chapter 3

Looking at Stars

The stargazers of the ancient world would, to begin with, have been as confused as novice stargazers today by the apparent chaos of stars that shine in the night sky. But they soon began to bring some kind of order to bear when they could see that the brightest stars formed patterns that they could recognize night after night. These star patterns provided a means of guiding them across the heavens.

Astronomers still use these star patterns, which we call the constellations, for the same purpose. Ancient astronomers named the constellations after figures suggested by the shapes of the star patterns, such as a crouching lion, a flying swan, and a scorpion poised ready to sting with its deadly tail. Indeed, these particular star patterns do passably resemble these animals. But the star patterns of most constellations look nothing like the figures they are named after.

Today we split up the heavens into 88 constellations. Many have been handed down to us from the ancient world, particularly from the Greeks. And we name them using the Latin equivalent of the Greek names. So the constellation that looks like a lion is named Leo, the swan Cygnus, and the scorpion Scorpius. Latin names are also used for the constellations introduced since classical times, mostly in the sixteenth and seventeenth centuries. Interestingly, they include objects that reflect some of the new inventions of the time, such as the telescope, microscope, and sextant.

Within each constellation, astronomers need to identify the stars. The standard method of doing so is to use letters of the Greek alphabet to grade the stars in order of their brightness. The brightest star is designated alpha (α), the next brightest beta (β), and so on.

A star in a particular constellation is named using the appropriate letter of the Greek alphabet and the genitive form of the Latin name of the constellation. So the brightest star in Leo is named Alpha Leonis; the second brightest star in Cygnus, Beta Cygni. Many of the brightest stars also have proper names, for example, Alpha Leonis is called Regulus and Beta Cygni, Albireo.

Great Gaseous Globes

Considering how far away the stars are—trillions upon trillions of miles—we know a great deal about them. This is primarily due to the fact that we have a star right on our celestial doorstep, as it were. This is the star we call the Sun. We gain further information by analyzing the faint light that reaches us from the stars.

Although stars vary individually in many ways, they are all basically similar. They are great globes of searing

hot gas that generate fantastic energy and pour it out into space as light, heat, and other kinds of radiation (rays), such as X-rays and radio waves. All these different kinds of radiation are intrinsically the same—they are electric and magnetic disturbances that travel in the form of waves, which we call electromagnetic waves. They differ in how long the wave is: X-rays have some of the shortest wavelengths, radio waves the longest (see page 22).

The most common chemical elements in stars are hydrogen and helium. There are also traces of as many as 70 other chemical elements, such as carbon, magnesium, and iron. On Earth these last three elements are solids, but in a star they and all the other elements exist as a kind of matter not found on Earth, called plasma. In the plasma state, elements are present not as atoms but as electrically charged particles immersed in a sea of electrons.

Producing Energy

The surface temperature of a star is typically thousands of degrees. But in the star's center temperatures reach tens of millions of degrees. Pressures in the center also reach millions of times atmospheric pressure on Earth. Under these extreme conditions, reactions take place that produce the

Dying Dwarf

The star at the center of this colorful gas cloud is one of the smallest stars, called a white dwarf. It represents a late stage in the life of a Sun-like star. It is tiny—not much bigger than Earth—but is incredibly dense and incredibly hot. The gas cloud, called the Little Ghost Nebula, is an example of one of the most colorful types of astronomical objects— a planetary nebula.

Distant Suns

Sparkling all colors of the rainbow, the distant suns we call the stars pour light and heat into the abysmal darkness of space. By analyzing the rainbowlike spectrum of the faint light that reaches us from the stars, astronomers can tell what they are made of, how fast they are traveling, and much more.

enormous energy that keeps the star shining, often for tens of billions of years.

The reactions taking place in a star's interior are nuclear fusion reactions, in which the nuclei (centers) of hydrogen atoms fuse together to form a new element—helium. During this process, a certain amount of mass is converted into energy according to Albert Einstein's famous equation $E=mc^2$ where E is the energy produced when a mass m is converted, and c is the speed of light. Since c^2 is a huge number, the conversion of even tiny amounts of mass produces huge amounts of energy.

Vive la Difference
Although all are basically similar, stars differ individually in many ways. For one thing, they vary greatly in size. The Sun, with a diameter of just under a million miles (around 1.5 million km), is a relatively small star, and there are many other stars of similar dimensions. But

there are also dwarf stars that are hundreds of times smaller than the Sun and giant stars that are tens of times bigger.

The biggest stars, called supergiants, are tens of times bigger still. For example, the supergiant Betelgeuse—the prominent alpha star in the constellation Orion—has a diameter of at least 250 million miles (400 million km). If it were located where the Sun is in the Solar System, it would stretch way beyond Earth.

Stars also differ widely in surface temperature. The Sun is a medium-hot star, with a surface temperature of about 10,000°F (5,500°C). The coolest stars have only about half this temperature, while the hottest have temperatures exceeding 72,000°F (40,000°C).

The temperature of a star's surface directly affects its color. The hottest stars appear blue–white, medium–hot stars such as the Sun appear yellow, and the coolest stars red.

Star Brightness
From even a quick glance at the night sky, we can see that not all stars have the same brightness. Some shine like beacons, while others can scarcely be seen. Astronomers grade the brightness of stars on a scale of magnitude, grading the brightest at 1st magnitude and the dimmest at 6th magnitude, with others graded in between.

A few exceptionally bright stars are given negative magnitudes. The brightest star in the sky, Sirius, for example, has a magnitude of –1.45. And stars too faint to be seen with the naked eye, but visible in telescopes, are given magnitudes greater than 6.

However, these magnitudes reflect only how bright the stars appear to be to our eyes. Because stars lie at different distances from us, these apparent magnitudes do not relate to their true brightness. So astronomers have devised a scale of true brightness, or absolute magnitude, that compares the brightness of stars when viewed from the same distance (33 light-years, or 10 parsecs).

Grouping Together

The stars in the constellations do not seem to change their positions year by year, or even century by century. The constellation patterns we see today are virtually identical to the ones the Greeks saw over 2,000 years ago. It seems as if the stars in the constellations are fixed and are grouped together in space. But nothing could be farther from the truth.

Virtually all the stars in a constellation lie different distances away. They appear to be grouped together only because they happen to lie in the same direction in space as we view them. And they are traveling independently through space in different directions. We can't see most of them move because they lie so far away.

The Sun, too, travels through space independently. But many other stars don't. They travel along with one or more companion stars. Most commonly, stars travel in pairs, forming what is called a binary star system. As they travel, they circle around each other, bound together by their mutual gravity.

Some binary systems are special, with the two stars circling each other in our line of sight. Because of this, each one periodically passes in front of, or eclipses, the other, leading to a sudden drop in brightness of the two-star system. We call it an eclipsing binary. It is one kind of variable star. Other variable stars vary in brightness because of changes going on inside them.

Star Clusters

When stars are born in the great dark clouds that permeate space, they are not born on their own (see page 33). Usually they are born in clusters of tens and even hundreds. We find many such clusters in the heavens. We call them open clusters, because the stars in them are only loosely packed together.

One of the best-known open clusters is to be found in the constellation Taurus. It is the Pleiades, also called the Seven Sisters because especially keen-sighted people might be able to make out its seven brightest stars. The Pleiades is a relatively young cluster, made up of hot blue stars. Similar open star clusters have also been spotted in other galaxies, confirming that they are a common product of star formation.

Open clusters typically consist of a few hundred stars. But there

Hot Clusters

This cluster of hot stars in our neighboring galaxy, the Large Magellanic Cloud, was born only a few million years ago. The intense radiation emitted by its stars is lighting up the hydrogen gas in the surrounding space, which glows typically red.

are clusters in the sky that contain stars in the hundreds of thousands. The stars congregate into a globe-shaped mass and are appropriately called globular clusters. Two spectacular globulars are visible to the naked eye in the Southern Hemisphere— Omega Centauri in the constellation Centaurus and 47 Tucanae in Tucana. There are another two fine, though fainter, globulars in the Northern Hemisphere—M3 in Canes Venatici and M13 in Hercules.

Open clusters are to be found on the spiral arms of the star system, or galaxy, to which all the stars in the sky belong (see page 78). Many of them lie quite close to us. But globular clusters all lie much farther away. They are found near the center of the galaxy, where they circle around the central bulge of stars there.

Between the Stars

Along with many different kinds of stars and clusters, the night sky dazzles with glowing clouds we call nebulae. They are made up of tenuous mixtures of gas and dust. Traces of gas and dust are found nearly everywhere between the stars, but only in places do they become dense enough to be detected.

The chief chemical elements that make up the gaseous part of nebulae are hydrogen and helium, but there are also significant amounts of oxygen, nitrogen, and iron. Chemical compounds—two or more elements combined together—are also present. Among them are water, alcohol, and ammonia. Finely divided grains of carbon make up most of the dust nebulae contain.

The most prominent nebula visible to the naked eye is found in the constellation Orion, visible as a misty patch south of the three stars that make up Orion's "belt." The bright-shining Orion Nebula is classed as an emission nebula because its gas particles glow, that is, emit light, when they are excited (energized) by the radiation from hot stars embedded within them. We also find bright emission nebulae in other galaxies, among the most spectacular being the Tarantula Nebula in our neighboring galaxy, the Large Magellanic Cloud.

Interstellar Ocean

Turbulent masses of gas, like waves on a stormy ocean, fill the space between stars in the constellation Sagittarius. This image shows part of the vast gas cloud known as the Omega or Swan Nebula. The wavelike patterns in the cloud have been sculpted and lit up by intense ultraviolet radiation from nearby massive new stars. Yellow-orange colors show the warmed surface of the dense hydrogen gas cloud. Blues and greens show less dense but hotter regions.

In the Dark

Often in interstellar space, concentrations of gas and dust accumulate but do not get lit up by embedded stars. Most of these dark nebulae remain invisible, but some we can detect. We can see some when they blot out the light from stars behind them. We can see others when they are silhouetted against the glowing gas of a bright nebula.

In the far southern constellation Crux, a prominent dark nebula looks like a hole in the bright expanse of the Milky Way. It is called the Coal Sack. In Orion, outlined by a bright emission nebula, lies the best-known dark nebula of them all, a dark mass resembling a horse's head.

The Birthplace of Stars

It is in the darkest and densest of interstellar gas and dust clouds that stars are born. We call them dark molecular clouds because the chemical elements present, mostly hydrogen, are present as molecules— stable combinations of atoms.

Dark molecular clouds can measure hundreds of light-years across and can remain quiescent for millions of years until something disturbs them—maybe the shock waves from the catastrophic explosion of a nearby star. When such a disturbance does occur, it forces regions of the cloud to become denser still. Now gravity comes into play, and the molecules in the regions start attracting one another. Regions in the cloud start to collapse.

Matter in the cloud is now splitting up into perhaps hundreds of collapsing clumps. As collapse proceeds, the clumps get denser. They also get hotter as the gravitational energy that the condensing matter releases is transformed into heat. The densest and hottest region in each clump is the center, or core. Temperatures there begin to reach millions of degrees, and pressures build up too. The core is well on the way to becoming a star. At this stage we call it a protostar.

Eventually, as more and more matter collapses onto the core and generates more heat, temperatures reach tens of millions of degrees. Now they trigger off nuclear fusion reactions between the nuclei of hydrogen atoms that release fantastic energy (see page 30). This energy makes its way to the surface of the erstwhile protostar and pours out into space as heat and light. The protostar is a protostar no longer but a new addition to the stellar universe.

Life and Death
Every newborn star takes a while to settle down into a stable state, but then begins to shine steadily. We might be forgiven, on seeing the apparently never-changing panoply of stars in the night sky, for thinking that stars live forever. But they most assuredly do not. Rather like living things on Earth, stars are born, mature, reach old age, and eventually die. But their lifespan is measured not in a few hundreds or thousands of years, but in millions and often billions of years.

How long a star lives depends mostly on its mass. Small stars with a low mass live the longest, with life spans of tens of billions of years. The relatively small star we know as the Sun should live for about 10 billion years—at present it is about halfway through its predicted lifespan. The most massive stars, however, lead much shorter lives. They may expire after just a few million years.

The Death of Sunlike Stars
The mass of a star determines not only when it dies, but also how. A relatively small star such as the Sun stops shining steadily when it runs out of the hydrogen "fuel" in its core that it has been using since it was born. The core then collapses, releasing energy that makes the outer layers of the star balloon out. The

Keyhole Nebula
The billowing clouds of interstellar gas and dust in the Keyhole Nebula in the constellation Carina are the birthplace of stars in their thousands. Cold, dark regions within the nebula will eventually collapse under gravity and spawn a host of new stars.

star expands maybe as much as 50 times to become a body we call a red giant—red because its surface is now very much cooler.

Eventually, the core of the giant star shrinks smaller still, becoming a superdense body called a white dwarf, not much bigger than Earth. Meanwhile, the outer layers of the star have been puffed off into space, creating one of the most beautiful of all astronomical creations—a planetary nebula.

The Death of Massive Stars
A star with a much greater mass than the Sun races through its life, shining with spectacular brilliance. It swells up when it begins to die. But because it is so massive to start with, it swells up to enormous proportions to become a supergiant, hundreds of times bigger across than the Sun.

But it does not remain a supergiant for long, maybe for only a few thousand years. Then its central core collapses in on itself violently, releasing so much energy

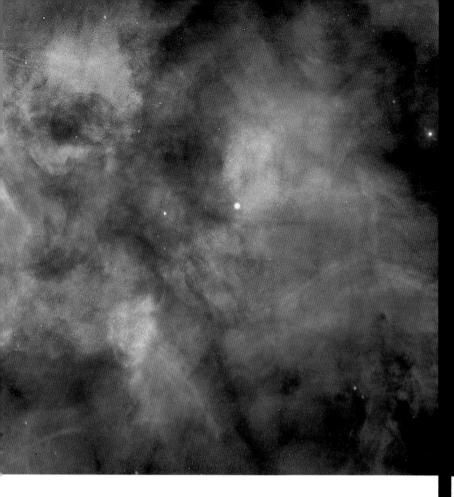

Identifying Nebulae and Clusters

Various methods are used to identify the different kinds of objects astronomers see in the heavens, such as nebulae and star clusters. The most familiar ones have proper names, which may describe, for example, in which constellation they are located—the Orion Nebula is one of the showpieces of the constellation Orion. Alternatively, the object may have a name that describes what it looks like. For example, the Eagle Nebula in the constellation Serpens is so named because it looks somewhat like a bird in flight.

that it triggers off an explosion that blows the star to bits. We call this catastrophic explosion a supernova.

Pulsars and Black Holes

If the mass of the collapsing core of the star is up to about three times the mass of the Sun, it becomes a tiny body made up of subatomic particles called neutrons, squashed tightly together. Typically less than about 12 miles (20 km) across, we call it a neutron star. Rapidly spinning, it releases pulses of energy like a celestial lighthouse. When we can detect its pulses, we call it a pulsar.

If, on the other hand, the mass of the collapsing star from a supernova explosion has a mass more than three times the mass of the Sun, the collapse continues beyond the neutron-star stage. The matter in the core is virtually crushed into nothing. All that remains is a region of space dominated by the gravity of the former core matter. The gravity is so great that the region of space will swallow up everything that comes nearby. Not even light can escape its clutches, which is why we call it a black hole. Obviously we can't see black holes, but we can infer their presence from X-radiation given out when superhot matter swirls into them.

M Nebulae and clusters may also be identified by an M number—the Eagle Nebula is also known as M16. M stands for Messier. Charles Messier was a French astronomer of the eighteenth century who was a fanatic hunter of comets. He got thoroughly fed up with mistaking the misty patches of nebulae and star clusters with what might be comets. So in 1781, he drew up a list of the most prominent nebulae and clusters and where they were located in the sky, so that he—and other comet hunters—would not confuse them with comets. Messier published his list in 1781. The Eagle Nebula was number 16 on his list, hence M16.

NGC There is yet another common way in which nebulae and clusters are identified, by an NGC number—the Eagle Nebula is NGC 6611. This number has its listing in the *New General Catalogue of Nebulae and Clusters of Stars*, the original version of which was prepared by the English astronomer John Dreyer and published in 1888. It was much more comprehensive than Messier's list (which eventually listed just 109 objects) and reflected the expansion of astronomical knowledge since Messier's time.

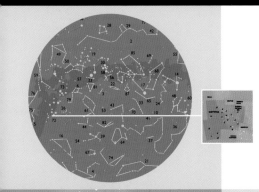

Hemisphere
South
Constellation
Sagittarius
(the Archer)
Distance
10,000 ly
Object type
Star Cloud
Object name
Sagittarius Star
Cloud
Image source
Hubble Space
Telescope

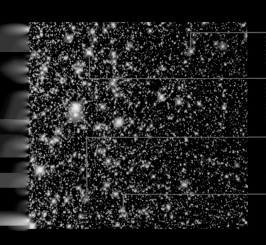

The brightest stars lie
much closer to us than
the Star Cloud

Young blue stars with
a mass and surface
temperature many
times those of the Sun

Older red stars are
mostly gigantic red
giants, very large but
cool

Dense swathe of stars,
tens of thousands of
light-years deep

Sparkling Starscape
*Studded with stars glittering like jewels on black
velvet, the Sagittarius Star Cloud is breathtakingly
beautiful when viewed by the Hubble Space Telescope.
When we look at it, we are peering into the very
heart of our Galaxy.*

Whenever you look at the night sky for a
while, you notice that the stars are not all the
same. They vary, for example, in brightness,
from very faint ones you can hardly see to
very bright ones that shine like celestial
beacons. And, if you look carefully, you notice
that the stars also vary in color. Some are
bluish white, some yellow, some orange, and
others red.

The differing colors of the stars are well seen
in this stunning image of the Sagittarius Star
Cloud. It is a particularly dense region of the
hazy white band we see in the sky that we
call the Milky Way. As its name implies, it is
located in one of the Southern Hemisphere's
most spectacular constellations—Sagittarius.
The Milky Way represents a "slice" through
our star system, or Galaxy, which we also call
the Milky Way Galaxy (see page 78).

The color of a star reveals its surface
temperature, one of its most important
"vital statistics." Most blue stars are young
and hot, with a surface temperature as
high as 72,000°F (40,000°C)—seven times
that of the Sun. They consume their nuclear
"fuel" quickly and live relatively short lives.

The coolest stars are red, with a typical
surface temperature of about 5,500°F
(3,000°C), around half that of the Sun. Some
of these cool, red stars are red giants, which
are stars originally like the Sun that are
nearing the end of their lives. Others are red
dwarf stars, with a small size and mass. They
are the stars that live longest—for tens of
billions of years.

AB7

One of the hottest stars ever found lies in the heart of this glowing cloud, or nebula, in one of our neighboring galaxies, the Small Magellanic Cloud. This galaxy is a smaller and somewhat more distant companion to the Large Magellanic Cloud. Both clouds are visible to the naked eye in far southern skies.

The star in question, AB7, is actually a double star, a type we call a binary. It consists of two stars that circle around each other in space, bound together by their mutual gravity.

The hottest and most massive star of the binary pair is a Wolf–Rayet star (named for the astronomers who discovered the type, C. J. E. Wolf and G. A. P. Rayet). Wolf–Rayet stars are noted for their very high surface temperatures and short lives. This one is estimated to have a temperature exceeding 215,000°F (120,000°C), around 20 times that of the Sun. Its companion is also massive, but about three times cooler.

Just as the Sun gives off streams of particles in a "solar wind," so do other stars, as "stellar winds." The binary pair in AB7 give off exceptionally strong stellar winds—a billion times more powerful than the solar wind.

As these winds blow out into the surrounding space, they exert enormous pressure on the interstellar matter—the matter that exists between the stars. They blow up this matter into huge bubbles and also "excite," or energize it, splitting up the atoms into ions, or electrically charged particles. As these ions lose energy, they give off visible radiation, which we see as different colors.

Hot Stuff

A technicolor cloud, or nebula, of glowing matter surrounds AB7, one of the hottest stars we know. The colors of the nebula reflect its composition. Seen in three dimensions, it would appear as a huge incandescent bubble in space.

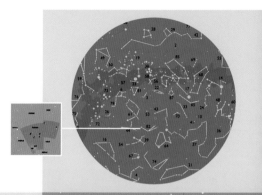

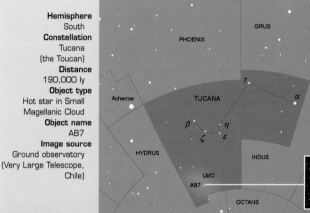

Hemisphere
South
Constellation
Tucana
(the Toucan)
Distance
190,000 ly
Object type
Hot star in Small Magellanic Cloud
Object name
AB7
Image source
Ground observatory
(Very Large Telescope, Chile)

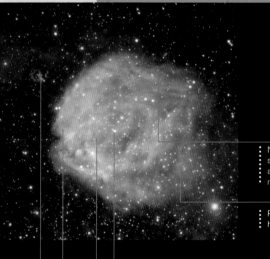

Nebula created when stellar winds come up against interstellar matter

Red light from hydrogen ions

Binary star AB7, source of strong stellar winds

Blue light from helium ions

Green filaments pinpoint site of recent supernova explosion

Green light from oxygen ions

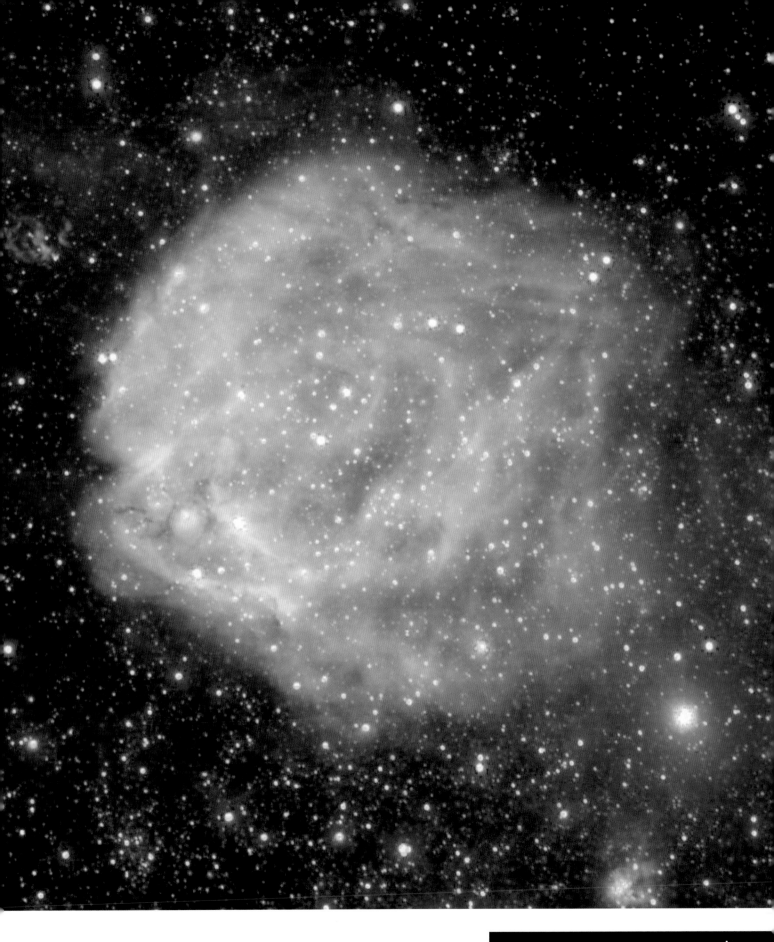

Pleiades M45 : Seven Sisters

The most outstanding feature of the constellation Taurus is the cluster of stars we know as the Pleiades, or Seven Sisters. Together, they have a combined brightness approaching the 1st magnitude, and therefore rivaling that of the constellation's brightest star, Aldebaran.

Brilliant in December and January skies, the Pleiades is easily located by following a line through the three stars that form Orion's Belt and Aldebaran. Very keen-sighted people may be able to make out the seven brightest stars in the cluster, hence its alternative name, the Seven Sisters.

The Pleiades is an open cluster, meaning that the stars in the group are relatively far apart (in contrast with the closely packed stars of a globular cluster).

In all, the cluster is made up of about 100 stars. They were born together out of the same dark molecular cloud. They are all blue, hot, and young—only about 78 million years old. They are surrounded by nebulosity—little clouds of interstellar matter lit up by the intense radiation from the stars. Although the Pleiades are at present grouped together, eventually they will drift apart, as happens with all open clusters.

Another feature of this particular image is the presence near the left edge of another bright "star." But star it is not. It is another very famous astronomical body—Halley's Comet. It passed close to the Pleiades in the spring of 1986, the last time it appeared in our skies.

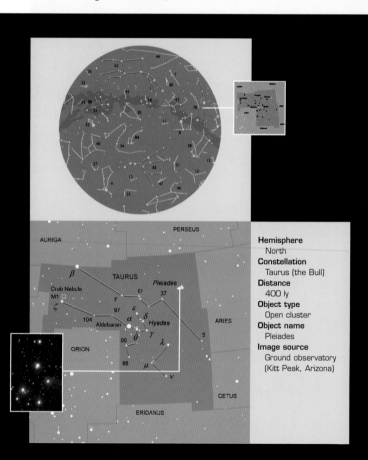

Hemisphere
  North
Constellation
  Taurus (the Bull)
Distance
  400 ly
Object type
  Open cluster
Object name
  Pleiades
Image source
  Ground observatory
  (Kitt Peak, Arizona)

The Seven Sisters

One of the finest sights in the heavens, the Pleiades, or Seven Sisters. The stars are named for characters in Greek mythology. The sisters known as the Pleiades were the beautiful daughters of the Greek hero Atlas and the nymph Pleione.

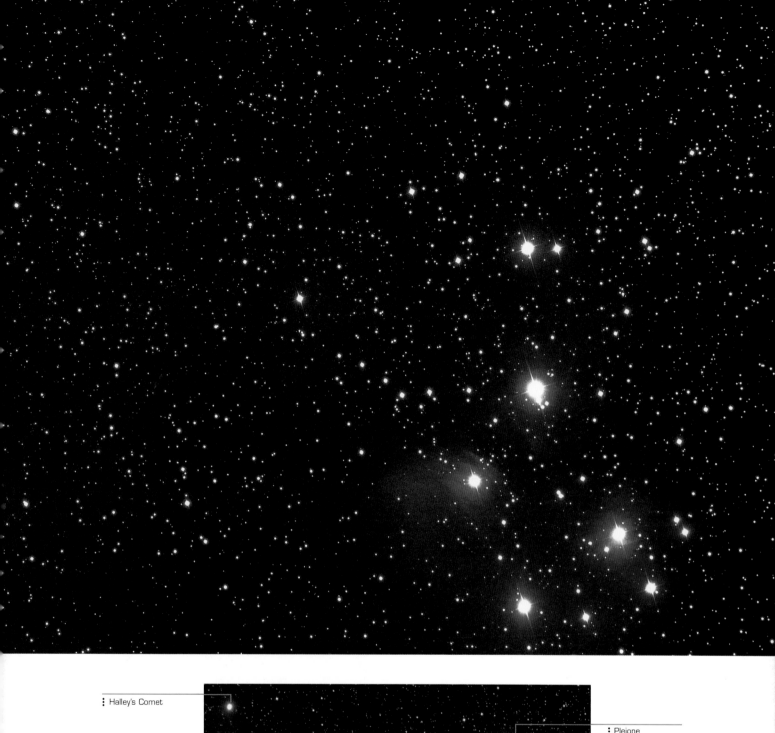

Halley's Comet

Atlas

Merope

Electra

Pleione

Alcyone

Maia

Taygeta

Asterope

Celaeno

Hodge 301

With its supersharp "eyes," the Hubble Space Telescope can picture starscapes in other galaxies as clearly as other telescopes picture starscapes in our own.

Here we see a star cluster known as Hodge 301, which resides within our galactic neighbor in the universe, the Large Magellanic Cloud (LMC). It lies within the Tarantula Nebula, a vast star-forming region that can be seen from the Earth with the naked eye. Within the nebula, stars are being born all the time. Most are not born singly, but in clusters of up to a few hundred.

The Hodge 301 cluster contains an abundance of brilliant massive stars, much larger and heavier than the Sun. These stars were born only a few million years ago, and they are racing through their life, consuming their hydrogen fuel voraciously. Some have already reached old age and have ballooned into gigantic red supergiants with diameters hundreds of millions of miles across.

These aging stars are on the brink of extinction, nearly ready to blast themselves to pieces in the most powerful "bang" in the universe, a supernova explosion. In a supernova, the core of the star collapses, and its outer layers are ejected into the surrounding space, accompanied by powerful shock waves.

Indeed, some of the stars in the Hodge 301 cluster have already exploded, sending shock waves rippling through the nebula. As the shock waves and ejecta (ejected matter) plow into the surrounding gas, they compress it into a multitude of denser sheets and filaments. The energy released excites the gas (gives it extra energy), which it then releases as radiation.

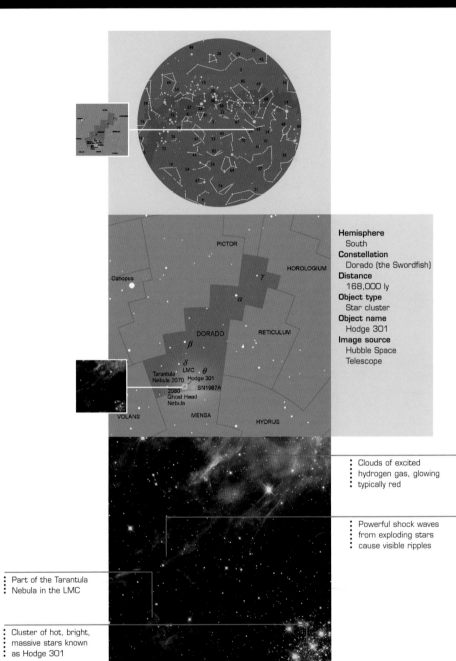

Hemisphere
South
Constellation
Dorado (the Swordfish)
Distance
168,000 ly
Object type
Star cluster
Object name
Hodge 301
Image source
Hubble Space Telescope

Clouds of excited hydrogen gas, glowing typically red

Powerful shock waves from exploding stars cause visible ripples

Part of the Tarantula Nebula in the LMC

Cluster of hot, bright, massive stars known as Hodge 301

Tarantula Nebula

A fascinating tapestry of stars and glowing gas in the Tarantula Nebula, revealed in exquisite closeup by the Hubble Space Telescope. At bottom right is the dense cluster of brilliant heavyweight stars that is Hodge 301. Other stars in the cluster have exploded, generating powerful shock waves that have created delicate ripples in the surrounding gas.

Hodge 301 | 43

M80 : NGC 6093

Scorpius is one of the most spectacular constellations in the heavens. Not only does it somewhat resemble the scorpion it is named for, but it also contains a host of fine astronomical objects. One is its brightest star, the noticeably reddish Antares, meaning "rival of Mars." And close by is M4, a globular cluster just visible to the naked eye under ideal conditions.

The image we see here is of another fine globular cluster, M80, located north of M4 roughly halfway between Antares and Beta (ß) Scorpii. M80 is one of the densest of the 150 or so globular clusters we know in our Galaxy. It lies about 28,000 light-years away

and circles, like other globulars, around the central bulge of the Galaxy.

As this Hubble Space Telescope image shows, the majority of stars in the cluster are red. They are red giants—dying stars that have swelled up to enormous dimensions, with a diameter up to 50 times that of the Sun.

Detailed analysis of the stars in the densest part of M80 shows that it contains a number of hot, blue stars that are unusually young and more massive than the other cluster stars. These have been called "blue stragglers."

Astronomers think that a blue straggler forms when two ordinary cluster stars collide and merge together into a single massive new star. The energy released during the collision makes the star unusually hot and blue, mimicking the behavior of a normal massive newborn star.

All the hundreds of thousands of stars in the M80 cluster are very old. They were born more than 10 billion years ago, only a few billion years after the universe itself came into being in the Big Bang.

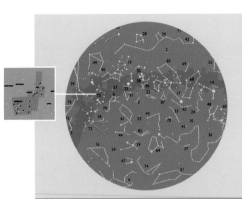

Most stars in the cluster are reddish and nearing the end of their lives

The densest part of M80 is in the center, where gravity is strongest

The cluster also contains much younger blue stars ("blue stragglers")

In the outer part of the cluster, the stars are more widely separated

Hemisphere
South
Constellation
Scorpius
(the Scorpion)
Distance
28,000 ly
Object type
Globular cluster
Object name
M80
Image source
Hubble Space Telescope

Glorious Globular

The globular cluster M80 is one of the many astronomical wonders in Scorpius. It is one of the densest globulars we know, a compact globe containing hundreds of thousands of stars. Most are very old, but a few, known as blue stragglers, are surprisingly young.

Chamaeleon I Complex

Billowing clouds of gas and dust occupy the space between the stars in the far southern constellation Chamaeleon. Radiation from embedded stars lights up the interstellar matter and creates the beautiful image we see here, taken by the Very Large Telescope in Chile. The whole region is known as the Chamaeleon I Complex.

The two brightest stars in the picture are very young and very hot. At this stage of their lives, they give off very intense radiation. The fine dust particles in the cloud of matter that surrounds them reflect the radiation, producing a blue color. This is typical of what is called a reflection nebula.

Adjacent to the bright star in the center of the picture is a dark region where few stars are visible. This is part of a vast dark molecular cloud, which is the main star-forming region in the Complex. The cloud is so dense that it is obscuring the light from background stars.

In the dark molecular cloud, matter exists as molecules at a temperature of about –430°F (–260°C). Star-formation begins when regions of the cloud become so dense that their matter starts to collapse under gravity. But a million years or more may go by until the cloud produces concentrated globes of matter that eventually light up to become new stars.

But not all these globes will make it that far. If they are too small, they never heat up enough to become fully-fledged stars. Instead, they end up, glowing feebly, as bodies known as brown dwarfs, often referred to as "failed stars." Many have been detected in the Chamaeleon Complex.

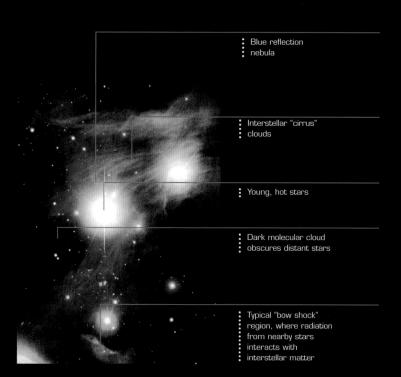

- Blue reflection nebula

- Interstellar "cirrus" clouds

- Young, hot stars

- Dark molecular cloud obscures distant stars

- Typical "bow shock" region, where radiation from nearby stars interacts with interstellar matter

Complex Chamaeleon

A vast star-forming region in the constellation Chamaeleon. Newborn stars are giving off powerful radiation that lights up surrounding interstellar matter. In places, this creates wispy nebulae that resemble cirrus clouds ("mares' tails") on Earth.

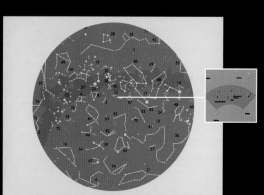

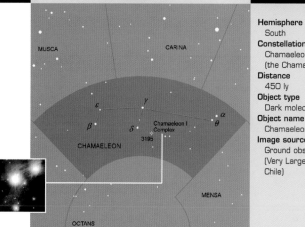

Hemisphere
South
Constellation
Chamaeleon
(the Chamaeleon)
Distance
450 ly
Object type
Dark molecular cloud
Object name
Chamaeleon I Complex
Image source
Ground observatory
(Very Large Telescope, Chile)

Carina Nebula : NGC 3372

Carina is an impressive southern constellation, partly embedded in the Milky Way. It boasts two outstanding stars. One is Canopus, which is the second-brightest star in the whole heavens, outshone only by Sirius, the Dog Star.

The other outstanding star is Eta (η) Carinae. It is one of the most massive stars we know, with a hundred times the mass of the Sun. Although at present it is only just visible to the naked eye, it has been much brighter in the past.

Astronomers classify Eta Carinae as an irregular, or eruptive variable, which flares up unpredictably. The last spectacular flare-up occurred in the 1840s, when it outshone Canopus and became the second-brightest star in the sky. Photographs of Eta Carinae taken by the Hubble Space Telescope show the vast clouds of gas and dust the star spewed out in that eruption.

The intense radiation from Eta Carinae pours out into space and sets the surrounding interstellar matter aglow. The result is the staggeringly beautiful Carina Nebula.

The Carina Nebula is divided by a V-shaped dark lane of dust, which seems to be the remnants of the original dark molecular cloud that collapsed to form stars a few million years ago. The nebula measures around 300 light-years across.

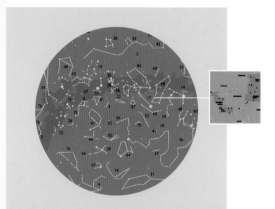

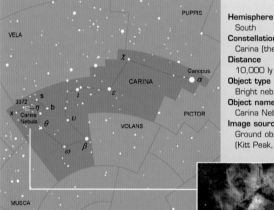

Hemisphere
South
Constellation
Carina (the Keel)
Distance
10,000 ly
Object type
Bright nebula
Object name
Carina Nebula
Image source
Ground observatory
(Kitt Peak, Arizona)

All Aglow

The interstellar gases surrounding Eta Carinae get excited by the star's intense radiation and glow, forming a classic bright nebula. It stands out against the starry backcloth of the Milky Way.

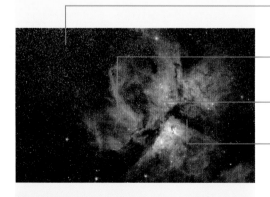

The dense mass of background stars that form the Milky Way

The Carina Nebula appears red, the color of excited hydrogen ions

This bright region hides the brilliant Eta Carinae

In these dark lanes, new stars are being born all the time

Orion Nebula M42

Straddling the celestial equator, the magnificent constellation Orion is equally visible to observers in both the Northern and Southern Hemispheres. Depicting the figure of a mighty hunter in Greek mythology, it is one of the most distinctive and easily recognized of all the constellations.

Three bright stars make up Orion's Belt, from which hangs his sword. Marking the sword handle is a misty patch, easily visible to the naked eye. This patch is a glowing mass of gas and dust, a bright nebula. Called the Orion Nebula or the Great Nebula, it is the nearest bright nebula to us, at a distance of about 1,600 light-years.

The Orion Nebula measures about 30 light-years across and is the glowing part of a vast dark star-forming region that pervades much of the constellation. It is lit up by a cluster containing thousands of hot, young stars, only about one million years old.

The brightest stars in the cluster form a multiple star complex called the Trapezium, because of their arrangement. They are difficult to see in visible light, but show up clearly at infrared wavelengths, as in this image.

The boundary where the bright nebula gives way to the darkness of the interstellar medium is often called the "bow-shock" area, or bright bar. It marks where the intense radiation and stellar winds coming from the stars of the Trapezium cluster are eating into the surrounding dark gas and dust. ➤

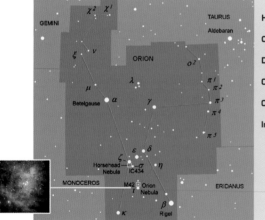

Hemisphere	South
Constellation	Orion
Distance	1,600 ly
Object type	Bright nebula
Object name	Orion Nebula
Image source	Ground observatory (Very Large Telescope, Chile)

: Other hot, young stars
: of the Trapezium cluster

: The four main stars of
: the Trapezium cluster

: The nebula is made to
: glow by the radiation
: from the stars in the
: cluster

: The bright bar, or
: bow-shock region

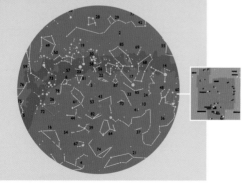

The Great Nebula

Orion's outstanding showpiece, the Orion Nebula. It is pictured here by the Very Large Telescope at Paranal, Chile, at infrared wavelengths, which reveal clearly the hot, young stars of the Trapezium cluster.

orion Nebula M42 | 51

Stars are being born
continually in the dark
cold expanses of gas
that permeate almost
all of the constellations

Hot, young, newborn
stars of the Trapezium
light up the Orion
Nebula

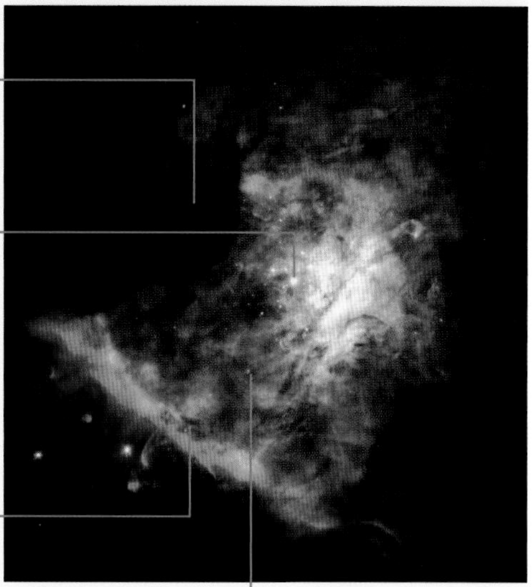

Dense globules of gas
condense, heat up, and
eventually spawn new
stars

Many, if not most
new stars become
surrounded by disks
of matter that may
condense into planets

Stellar Nursery

*This Hubble image shows extraordinary detail in the Orion
Nebula. The swirling gas clouds hide swarms of newborn stars.
Also visible are globules containing protostars, surrounded by
disks of gas and dust.*

➤ The Hubble Space Telescope has given
astronomers a new perspective on the
Orion Nebula. This mosaic image spans a
region of space around 2.5 light-years
across. It reveals turbulent clouds of gas
and dust tinged orange by glowing
hydrogen, creating a picture akin to the
angry clouds of a stormy sunset on Earth.

Clearly identified in the image are discrete
spherical and oval globules, which contain
protostars about to shine forth on the
universe. As a protostar lights up,

surrounding gas and dust form into a
swirling disk around it. Over time, matter
in the disk could condense to form
planets. That is why such disks are called
protoplanetary disks, or proplyds for
short. In all, more than 150 proplyds
have been identified in Hubble images of
the Orion Nebula.

The Orion Nebula is not the only spectacular nebula in the constellation Orion. Another is the Horsehead Nebula. But, whereas the Orion Nebula is readily visible to the naked eye, the Horsehead can be seen only in relatively large telescopes. This image of the Horsehead was taken by one of the four instruments that make up the Very Large Telescope in Chile.

The Horsehead Nebula is well named. It is a dark nebula whose shape bears an uncanny likeness to the head and mane of a horse (or rather a seahorse). It is thrown into dramatic silhouette by the bright emission nebula behind it, called IC 434.

The Horsehead is located in Orion close to Zeta Orionis, the most southerly of the three bright stars that form Orion's Belt. It forms part of Orion's northern molecular cloud, whereas the Orion Nebula forms part of the southern cloud. Both clouds are the densest star-forming regions in the constellation.

Astronomers often refer to fingers of dark gas and dust like the Horsehead as "elephant trunks." Other examples are the so-called "pillars of creation" in the Eagle Nebula (see page 58). All such structures are only temporary. The radiation creating the bright nebula moves into the dark cloud and sets that aglow too. So as time goes by, the Horsehead will change shape and eventually disappear—but not for several thousand years.

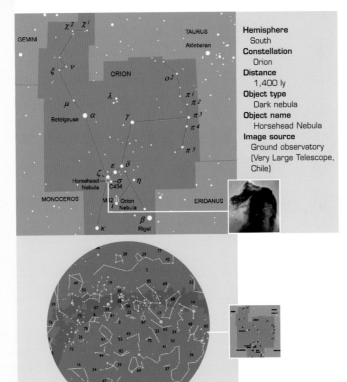

Hemisphere	South
Constellation	Orion
Distance	1,400 ly
Object type	Dark nebula
Object name	Horsehead Nebula
Image source	Ground observatory (Very Large Telescope, Chile)

Rearing its head

One of the most beloved of all astronomical images, and one that truly lives up to its name, the Horsehead Nebula in Orion. In practice, it is just a dusty mass, silhouetted against a bright nebula.

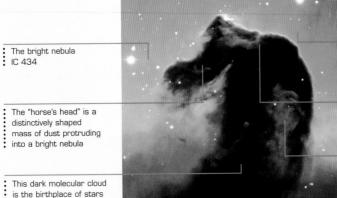

The bright nebula IC 434

The "horse's head" is a distinctively shaped mass of dust protruding into a bright nebula

This dark molecular cloud is the birthplace of stars

Red is the color of glowing hydrogen

Green shows scattered starlight

Brown shows where there is obscuring dust in the foreground

Eagle Nebula : M16

The Eagle Nebula is located in Serpens Cauda, the tail part of the constellation Serpens, which is unique among constellations because it is split in two (by Ophiuchus).

The nebula is a typical emission nebula, in which interstellar hydrogen is triggered into emitting its characteristic red glow by the intense ultraviolet radiation from an embedded cluster of hot, young stars.

Sometimes both nebula and cluster are referred to as M16, but often just the nebula itself is denoted M16 and the associated cluster NGC 6611. The stars in the cluster are only about 5 million years old, which is very young by stellar standards. (By comparison, the Sun is around 5 billion years old.)

The name of the nebula refers not to the shape of the nebula itself but to that of dust lanes at its center, which show up prominently in long-exposure photographs. These dust lanes are examples of features astronomers call "elephant trunks"—dark columns of gas and dust in which stars are forming.

On the following pages, the Hubble Space Telescope and the Very Large Telescope in Chile feature these columns in close-up, demonstrating vividly the wealth of extra detail that can be extracted using modern instruments and techniques. ➤

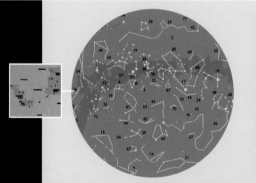

The Eagle's Lair

A fine view of the Eagle Nebula taken from a ground-based observatory. The nebula can be spotted in binoculars and small telescopes. It is best found by locating Gamma Scuti in the adjacent constellation and then scanning slightly northwest.

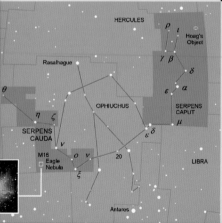

Hemisphere
South
Constellation
Serpens
(the Serpent)
Distance
6,500 ly
Object type
Bright nebula
Object name
Eagle Nebula
Scale
Ground observatory
(Kitt Peak, Arizona)

The dark features
whose shape gives the
Eagle Nebula its name

The glowing hydrogen
of the Eagle Nebula

The open star cluster
NGC 6611

The dense
background of stars
in the Milky Way

➤ It was on the first day of April 1995 that the Hubble Space Telescope returned one of the most dramatic astronomical images ever. It shows towering columns of dark gas and dust thrown into sharp relief as they are backlit by stars behind them.

Hubble scientists called these columns the "elephant trunks" in the center of the Eagle Nebula—the pillars of creation. They are the birthplace of new stars. The background radiation that illuminates the pillars comes from the star cluster NGC 6611 (out of the picture at top). It is also destroying them.

At the top and elsewhere on the pillars are small bumps or protrusions rather like fingers. Known as EGGs, or evaporating gaseous globules, they seem to show where newborn stars are emerging from the pillars as the background radiation evaporates the surrounding dense, dark gas. ➤

Amazing Columns

The pillars of creation, one of the best-known images ever returned by the Hubble Space Telescope. The remnants of a much bigger dark cloud, they will eventually disappear themselves.

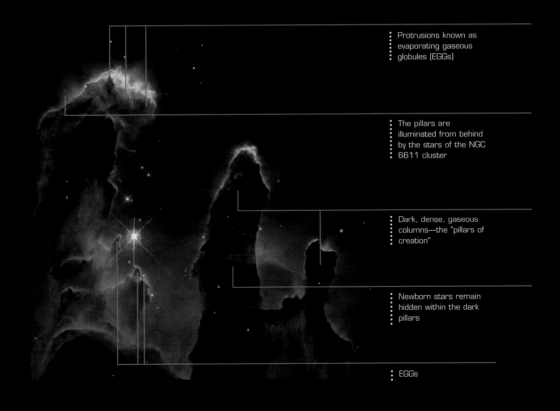

Protrusions known as evaporating gaseous globules (EGGs)

The pillars are illuminated from behind by the stars of the NGC 6611 cluster

Dark, dense, gaseous columns—the "pillars of creation"

Newborn stars remain hidden within the dark pillars

EGGs

➤ This alternative view of the Eagle Nebula's pillars of creation is dramatically different from the one returned by the Hubble Space Telescope. The image was taken by the Very Large Telescope in Chile at infrared wavelengths. It shows much more detail than the Hubble image, which was taken in visible light.

Infrared light readily penetrates clouds of gas and dust. This means that it can pass right through all except the densest parts of the famous pillars, revealing inside them many newborn stars.

On a broader scale in the image, infrared light has also passed through the layers of gas in the surrounding bright nebula. This allows us to see the myriad of background stars of the Milky Way.

The little purple arc pinpointed in the image is a fast-moving clump of hot gas coming from a very young star. This type of feature is called a Herbig-Haro object.

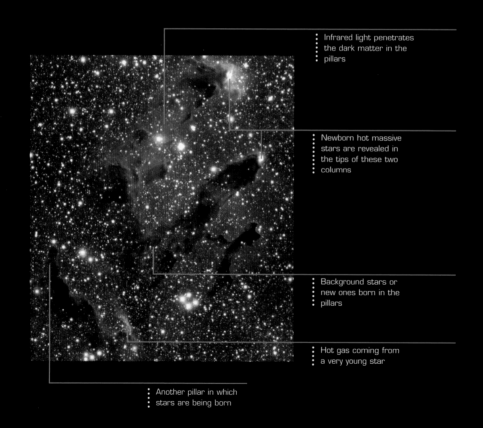

Infrared light penetrates the dark matter in the pillars

Newborn hot massive stars are revealed in the tips of these two columns

Background stars or new ones born in the pillars

Hot gas coming from a very young star

Another pillar in which stars are being born

Through the Pillars

View in infrared light of the central region of the Eagle Nebula and the famous pillars of creation, taken from the Very Large Telescope in Chile. Seen in this light, the pillars have become almost transparent, and stars show up in their thousands.

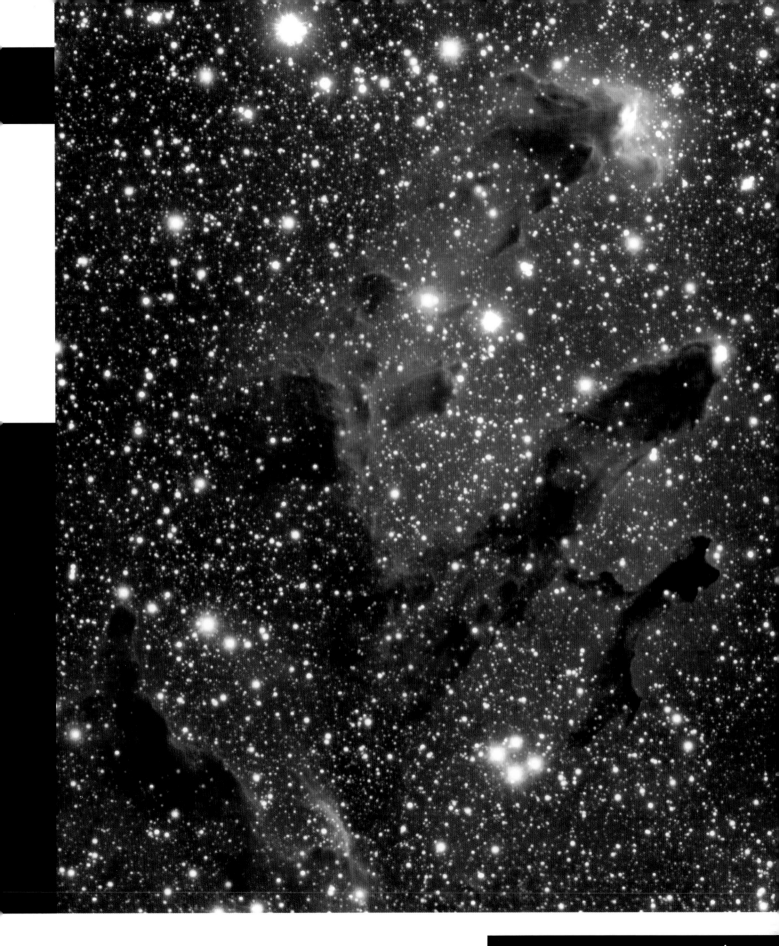

Eagle Nebula | 61

Helix Nebula : NGC 7293

Some 12,000 years ago, a Sunlike star in the constellation Aquarius was in its death throes. Earlier, it had expanded into a red giant, but was now puffing off its outer gaseous layers into space. The ejected gas streamed into space to form a colorful cloud, which we see today, still expanding, as the Helix Nebula or NGC 7293.

The Helix is a fine example of a planetary nebula. But such nebulae have nothing whatsoever to do with planets. Eighteenth-century astronomers called them this because, in the telescopes of the day, they showed up as a disk, rather like the planets do.

In photographs, the Helix Nebula looks as though it has some kind of coiled, or springlike structure ("helix" is an alternative name for coil). But astronomers now reckon that its structure is more complicated. They have come to this conclusion after studying the image shown here, which is based on combined data from the Hubble Space Telescope and a telescope at Kitt Peak Observatory in Arizona.

Studies have suggested that the Helix in fact consists of two disks nearly perpendicular to each other. We see the larger and more colorful outer disk head-on. The inner, more transparent disk is at right angles to it, coming, as it were, out of the page. This inner disk of matter was not ejected until about 6,000 years ago.

Visible around the inside of the ringlike structure of the Helix are tadpolelike features, with glowing heads and fainter tails. Scientists have named them cometary knots because of their resemblance to comets.

Exotic Flower

The Helix is among the most beautiful of all planetary nebulae, resembling an exotic tropical flower. Astronomers are finding that its structure is much more complex than was once thought.

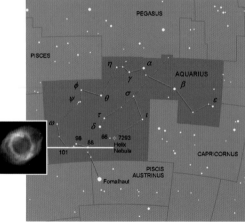

Hemisphere	South
Constellation	Aquarius (the Water-Bearer)
Distance	650 ly
Object type	Planetary nebula
Object name	Helix Nebula
Image source	Hubble Space Telescope

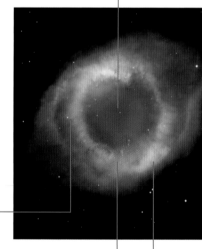

The dying star that ejected matter to form the Helix Nebula

Inner disk of ejected material

Tadpolelike "cometary knots"

Outer disk of ejected material

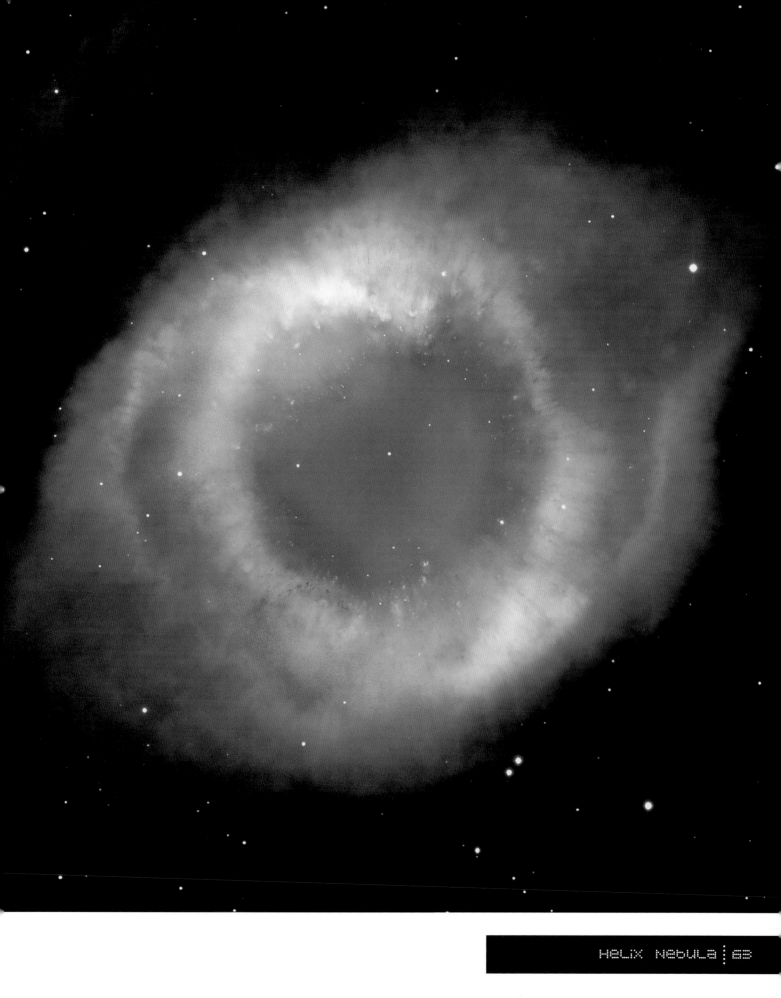

HELIX NEBULA | 65

Cat's-Eye Nebula

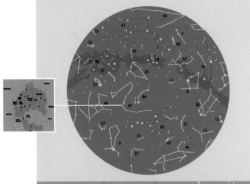

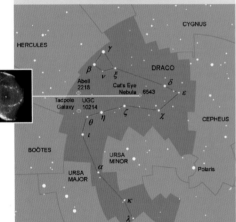

"Beautiful cosmic flowers in full bloom" is one way of describing some of the most visually stunning objects we know in the universe. They are planetary nebulae—nothing to do with planets, but the ghostly remnants of dying stars.

The Cat's-Eye Nebula in Draco is one of the most beautiful and most complex planetary nebulae we know. It is made up of a series of expanding gaseous shells that mark when material has been ejected from the dying star. This star—or rather what's left of it—lies in the center. Small and very hot, it is a white dwarf. Its temperature may be as high as 180,000°F (100,000°C). Its diameter is probably around 8,000 miles (12,000 km), making it about the same size as the Earth.

Compared with other planetary nebulae, the Cat's-Eye shows unusual features. For example, the shells of gas that surround the central star are off-center. This suggests to astronomers that the central star may in fact be a binary system, consisting of two stars circling around each other, close together. The dynamics of such a system could explain the off-center configuration.

Another oddity of the Cat's-Eye is the pair of blue and green arcs at the outer edge. It is thought that they might be caused by twin jets of gas being given off by the white dwarf's companion star. As the jets hit the surrounding interstellar medium at the edge of the expanding shells, they plow into it and make it glow.

The colors in the picture relate to the different kinds of atoms present. Red indicates the presence of hydrogen; blue, oxygen; and green, nitrogen.

Hemisphere
North
Constellation
Draco (the Dragon)
Distance
3,000 ly
Object type
Planetary nebula
Object name
Cat's-Eye Nebula
Image source
Hubble Space Telescope

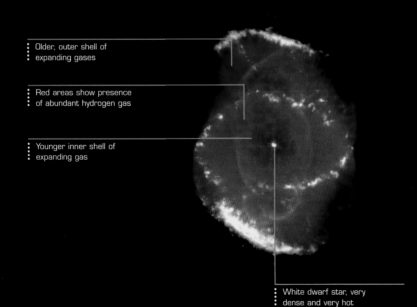

Older, outer shell of expanding gases

Red areas show presence of abundant hydrogen gas

Younger inner shell of expanding gas

White dwarf star, very dense and very hot

Cat's-Eye Nebula

This picture of the aptly named Cat's-Eye Nebula is one of the best-known Hubble images, which shows fascinating structure. Astronomers think that the asymmetric structure of the nebula could be explained if the central white dwarf star were part of a binary, two-star system.

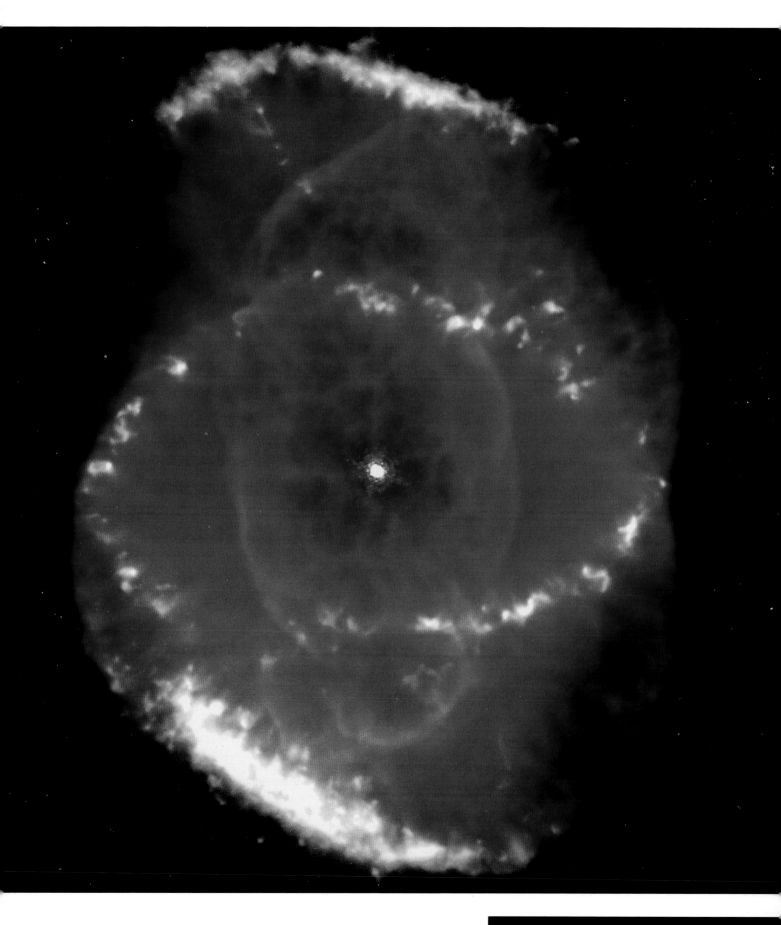

cat's-eye Nebula 68

V838 Monocerotis

Located some 20,000 light-years away in the constellation Monoceros is a red supergiant star known as V838 Monocerotis (V838 Mon). Like all supergiants, V838 Mon is highly unstable and in the near future will almost certainly blast itself to pieces as a supernova.

Early in 2002, the star suddenly brightened for several weeks. Two years later, the Hubble Space Telescope returned this remarkable image of V838 Mon and the surrounding region. It shows the star cocooned in a luminous sphere.

What it reveals is a so-called light echo, the reflection from a spherical cloud of gas and dust of light from the 2002 outburst. This material had been ejected by V838 Mon tens of thousands of years ago and remained invisible until suddenly lit up by the brilliant outburst of 2002.

This sudden outburst, during which V838 Mon flared up to become half a million times brighter than the Sun, had similarities to a class of objects called novae. Novae are faint stars that suddenly brighten because of violent explosions on their surface.

But V838 Mon showed quite different behavior from a nova—for example, remaining cool and red instead of becoming very hot and brilliant white. What happened was that it suddenly ballooned in size and brightness before returning to its former state. Astronomers reckon that V838 Mon's outburst may represent a transitory stage in stellar evolution that is rarely seen.

Starry Night

Hubble astronomers have likened this image of the expanding halo of light around V838 Mon to Vincent Van Gogh's famous painting called "Starry Night," which features bold whorls of light sweeping across a raging night sky.

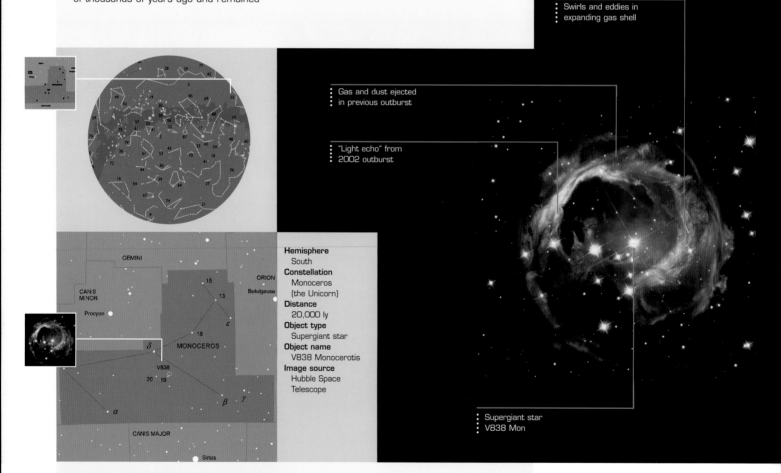

Swirls and eddies in expanding gas shell

Gas and dust ejected in previous outburst

"Light echo" from 2002 outburst

GEMINI

ORION

Betelgeuse

CANIS MINOR

Procyon

15

13

ε

18

δ

MONOCEROS

V838

20   19

β   γ

α

CANIS MAJOR

Sirius

Hemisphere
South
Constellation
Monoceros
(the Unicorn)
Distance
20,000 ly
Object type
Supergiant star
Object name
V838 Monocerotis
Image source
Hubble Space
Telescope

Supergiant star
V838 Mon

Supernova 1987A

On February 23, 1987, a bright new star appeared in our neighboring galaxy, the Large Magellanic Cloud. It was located close to the sprawling Tarantula Nebula, so named because of its spidery shape. In reality, it was not a new star but a faint one that had suddenly increased in brightness millions of times.

What had happened was that the star, known as Sanduleak -69°202, had blasted itself to pieces in a supernova explosion. Since it was the first supernova discovered in 1987, it was designated SN (supernova) 1987A.

Even though the catastrophic explosion had taken place some 160,000 light-years away, it was so bright that it was clearly visible from Earth with the naked eye. It reached its peak brightness in May 1987, when it reached a visual magnitude of 2.8. By then, it was estimated to be as bright as 250 million Suns.

SN 1987A was in fact the first supernova to be visible to the naked eye for 383 years. The last was the one witnessed by German astronomer Johannes Kepler in 1604.

The star that went supernova—Sanduleak -69°202—was a huge supergiant, with a mass at least 20 times that of the Sun. It had raced through its life cycle at lightning speed and was only about 20 million years old when it exploded.

Stars in the same neighborhood also appear to be massive stars that will eventually suffer the same fate. So will the cluster of stars that illuminates the Tarantula Nebula. Known as the 30 Doradus supercluster, it contains around a hundred stars with a mass up to 50 times that of the Sun. ➤

Stellar Suicide

The 1987A supernova rivals the Tarantula Nebula in brightness in this image taken in late February 1987. It became even more brilliant, easily visible to the naked eye to observers in the Southern Hemisphere.

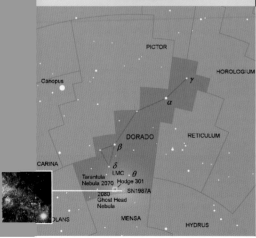

Hemisphere
South
Constellation
Dorado
(the Swordfish)
Distance
160,000 ly
Object type
Supernova
Object name
Supernova 1987A
Image source
Ground observatory
(Anglo-Australian
Telescope)

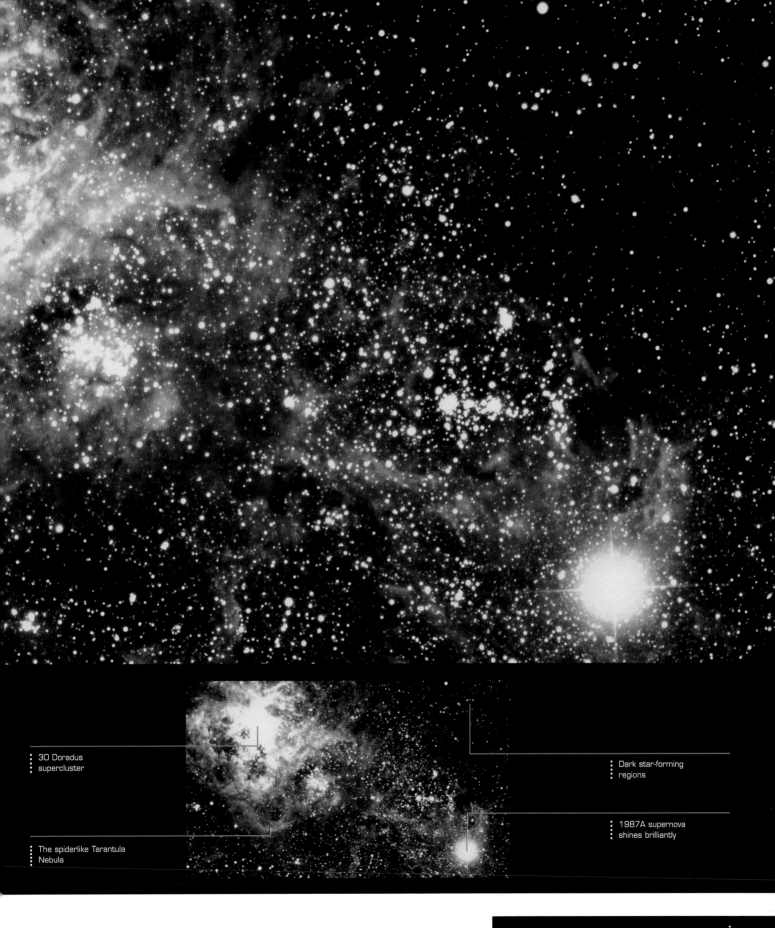

30 Doradus
supercluster

Dark star-forming
regions

1987A supernova
shines brilliantly

The spiderlike Tarantula
Nebula

The Hubble Space Telescope had not been launched when the 1987A supernova occurred in the Large Magellanic Cloud. But shortly after it had been launched, in 1990, it spotted a bright ring around the exploded star. Four years later, after it had been refurbished, it discovered two further rings.

In our view from Earth, all the rings seem to intersect. But in reality they are located in different planes. The bright central ring seems to be in the plane of the supernova. It is probably the "light echo" of the blast,

illuminating gas and dust that the doomed star had ejected earlier.

The other two fainter rings seem to lie in different planes, one in front of, and one behind, the 1987A explosion site. Astronomers are still not sure of the origin of these rings. One suggestion is that they are previously ejected material lit up by a high-energy beam of radiation from an invisible close companion of the star that went supernova.

Ring-a-Ring

A complex ring system now circles the site of the SN 1987A explosion, which is located near the edge of the Large Magellanic Cloud. The two bright stars apparently located in the outer rings are not associated with them.

Turbulent gas cloud in Tarantula Nebula

Star not associated with supernova

One of two large rings out of the plane

Site of 1987A supernova

Central ring of illuminated material in plane of supernova

Crab Nebula : M1

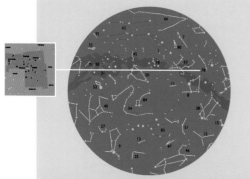

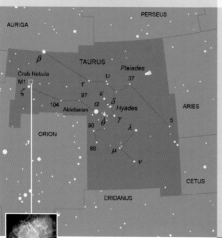

The ancient Chinese were skilled astronomers. Records show that they were observing comets from at least 1400 B.C. Later, in 1054 A.D., the Chinese recorded the appearance of a "guest star" in the heavens, in the constellation Taurus. It appeared suddenly and blazed so bright that it was visible to the naked eye, even during the daytime. It faded slowly as the months went by and then disappeared from view.

Today, we know that this "guest star" was a faint star that became millions of times more brilliant when it exploded as a supernova and blew itself to bits. The outer layers of the star were blasted into space, where they formed an ever-expanding cloud. Just the tiny dense core of the star remained.

Hemisphere
North
Constellation
Taurus (the Bull)
Distance
6,000 ly
Object type
Supernova remnant
Object name
Crab Nebula
Image source
Ground observatory (Very Large Telescope, Chile)

When the French astronomer Charles Messier began compiling his list of nebulae and star clusters in 1781, he spotted the cloud from the 1054 explosion and made it number 1 in his list (M1). Today, we call M1 the Crab Nebula, and class it as a supernova remnant.

The outer part of the nebula is the debris blasted into space during the supernova explosion. It shows up as intriguing colored filaments. In the middle of the colorful cloud is the collapsed core of the original star. It is a rapidly rotating neutron star that flashes beams of energy into space. Because we see these flashes, or pulses, of radiation, we call it the Crab pulsar.

The pulsar is associated with a powerful magnetic field, which whirls electrons around in the middle of the nebula to create the pale blue color we see in this image. ➤

Supernova remnant still expanding

Crab pulsar

Gas expanding in the form of delicate filaments

Red shows clouds of neutral hydrogen atoms

Blue shows radiation emitted by high-speed electrons

Green shows filaments of expanding hydrogen ions

Crab and Bull

A glorious picture of the Crab Nebula in Taurus (the Bull), taken by the Very Large Telescope in Chile. The nebula now measures some 15 light-years across, but it is still expanding at the phenomenal rate of nearly 1,200 miles (2,000 km) a second.

➤ The Hubble Space Telescope and the Chandra X-ray Observatory have both targeted the Crab Nebula, particularly the central region around the Crab pulsar. This image incorporates data returned from both observatories, so we are viewing the region in both X-rays and visible light.

The Crab pulsar is a rapidly rotating neutron star. It has a greater mass than the Sun, squeezed into a sphere only about 6 miles (10km) across. It spins around 30 times a second, beaming twin jets of radiation into space. It is difficult to imagine something the size of a small town spinning around that fast. But, compared with some pulsars, it is a slowcoach. The fastest we know spins around over 600 times a second.

In the image, the powerful radiation emitted by the pulsar creates rings as it comes up against surrounding gas. The inner ring, closest to the pulsar, shows where the most energetic particles are to be found. This ring gives off X-rays, detected by the Chandra Observatory. In the outer ring, the particles are less energetic and give off visible as well as X-rays. And this ring shows up in Hubble as well as Chandra images.

Another feature in the image is a jet of electrons moving in a direction perpendicular to the plane of the rings. It is made up of electrons, moving at half the speed of light.

Cosmic Heavyweight

The pulsar at the heart of the Crab Nebula is made up of neutrons packed solidly together. This is the densest form of matter we know. Just a teaspoonful of it would weigh millions of tons. It is a form of matter we call degenerate matter.

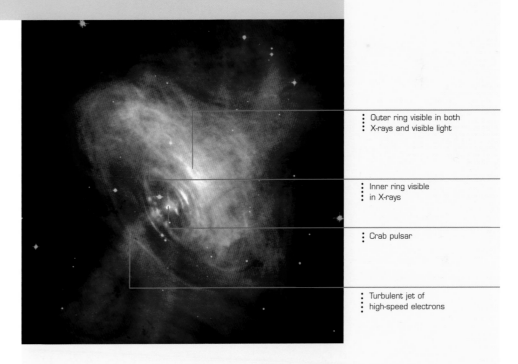

Outer ring visible in both X-rays and visible light

Inner ring visible in X-rays

Crab pulsar

Turbulent jet of high-speed electrons

Chapter 4

Looking at Galaxies

On any clear night, we can see about 2,500 stars above the horizon with the naked eye. Using binoculars, we can see many thousands more. Using a powerful telescope, we can see stars in their millions. These stars seem scattered haphazardly about in all directions, separated by the infinite blackness of empty space.

So, from our earthly perspective, we might be forgiven for thinking that the universe itself consists of stars scattered haphazardly about in empty space—but it doesn't. The stars in the universe congregate together into great star islands, and it is many star islands and the space between them that make up the universe. We call these star islands galaxies.

All the stars we see in the night sky belong to our own galaxy, which we usually call the Galaxy. If we could travel in a spaceship far out into space and look back, we would see that the Galaxy is shaped roughly like a disk with a bulge in the middle. The disk is not continuous but is formed of curved "arms" of stars that spiral out from the central bulge.

Our local star, the Sun, resides inside one of these spiral arms. And the stars around it are the ones that feature prominently in the night sky. In most directions in the sky, the stars are fairly well scattered. This is because we are, as it were, looking through the sides of the disk and through a relatively narrow thickness of stars.

But when we look along the plane of the disk, we are looking through a much greater thickness of stars, which extend all the way to each edge of the disk. This makes the stars in these directions seem to be packed closely together. In the sky, we see them as a hazy white band, which we call the Milky Way.

Only faint to the naked eye, but dazzling in binoculars, the Milky Way runs through many constellations, including Cassiopeia and Cygnus in the Northern Hemisphere and Scorpius and Sagittarius in the Southern. It is densest and brightest in Sagittarius because the center of the Galaxy lies in that direction.

The Galaxy

The Milky Way, the hazy band we see in the sky, represents a "slice" through our Galaxy, and therefore our Galaxy is also called the Milky Way Galaxy. It is huge, containing as many as 200 billion stars and vast clouds of gas and dust, some dark, some brilliant.

From one side to the other, the Galaxy measures around 100,000 light-years. The bulge at the center, or nucleus, is about 6,000 light-years across, while the disk averages only about 2,000 light-years. The Sun lies about 25,000 light-years from the center.

Like all the other galaxies, our Galaxy is rushing headlong through space, taking part in the general expansion of the universe. It also spins around, and from a distance would look like a flaming pinwheel firework.

The Spiral Arms

Even though we are located in the disk of the Galaxy, we can detect and map its spiral arms, using radio telescopes. It has two major arms and segments of several others. The two main arms are the Sagittarius and Perseus arms, named for the constellations they run through.

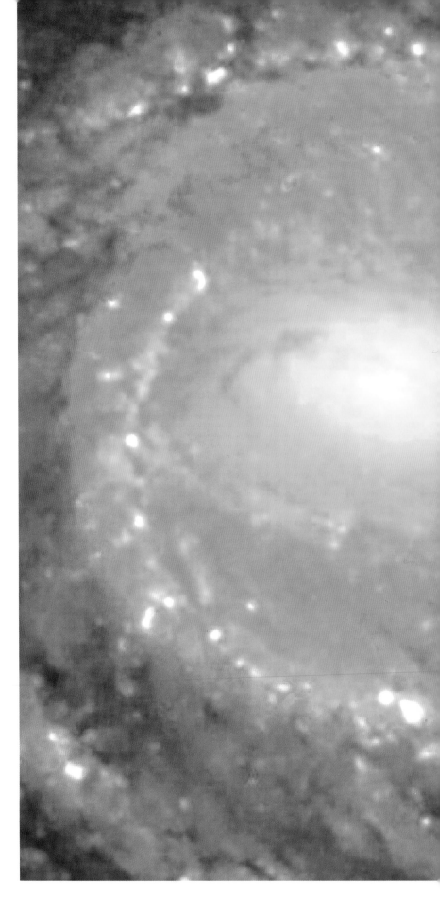

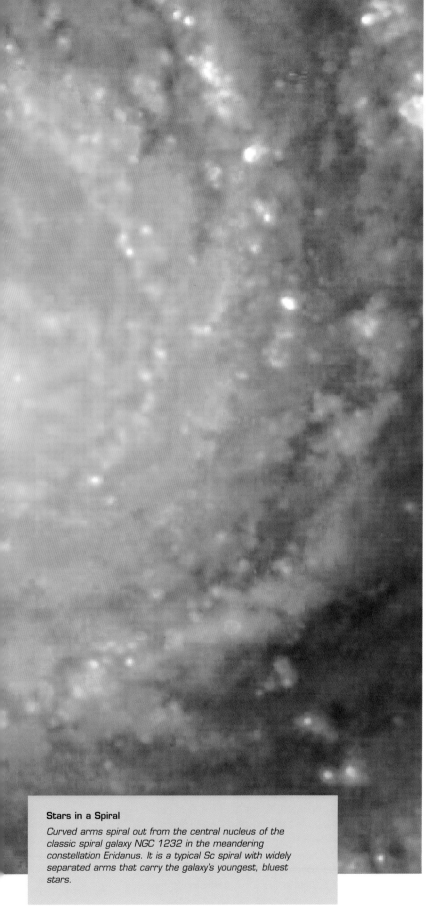

Stars in a Spiral

Curved arms spiral out from the central nucleus of the classic spiral galaxy NGC 1232 in the meandering constellation Eridanus. It is a typical Sc spiral with widely separated arms that carry the galaxy's youngest, bluest stars.

The Sagittarius Arm wraps itself almost all around the whole Galaxy. It carries some of the most spectacular nebulae and star-forming regions, including the Eagle, Trifid, and Lagoon nebulae. The Perseus Arm is more fragmented. Among the familiar objects it carries are the stunning Rosette Nebula and the Crab Nebula, the remnants of a star that exploded in 1054 A.D.

The Sun sits on the Orion or Local Arm, which forms a kind of bridge between the Sagittarius and Perseus Arms. Sitting on the arm also and located relatively close to us are some of the most outstanding objects in the sky, such as the amazing Orion Nebula, the Pleiades star cluster, and brilliant supergiant stars such as Betelgeuse, Rigel, and Canopus.

The Spirals

There are many other galaxies in the universe similar to our own Galaxy that have curved arms spiraling from the center. They are called spiral (S) galaxies. Another typical spiral is a relatively close neighbor of ours, the Andromeda galaxy.

Not all spirals have exactly the same shape. Galaxy pioneer Edwin Hubble classified them according to the openness of their spiral arms. Sa spiral galaxies have their arms relatively close together, while Sb and Sc galaxies have progressively more separated arms. Some spirals have a prominent bar of stars running through the nuclear bulge and are classed as barred-spiral (SB) galaxies. Again they are graded by the openness of their spiral arms into SBa, SBb, and SBc.

Other spirals have the similar mix of stars, nebulae, clusters, and so on, as our own Galaxy. The stars in the nuclear bulge are relatively old, while those in the spiral arms are relatively young, and it is in the spiral arms that most new star formation takes place.

Around the Bulge

Circling the nuclear bulge of galaxies are the globe-shaped concentration of stars we call globular clusters. They circle at some distance from the bulge and pursue independent orbits. They are among the oldest bodies in galaxies, probably forming first. Typically, they are about 10 billion years old.

Beyond the realm of the globular clusters, there appears to be a general "halo" of tenuous matter extending hundreds of thousands of light-years from the nucleus. Astronomers estimate that this could be made up of invisible, dark matter. The presence of dark matter in space is crucial in deciding how the universe might end (see page 15).

Ellipticals

As many as half of all galaxies are elliptical in shape. Some are completely spherical, but most have an oval shape. We still use the method Edwin Hubble developed to classify elliptical (E) galaxies. He graded them according to how flattened they were, from E0 (spherical) to E7 (most flattened).

Ellipticals differ primarily from spirals in their lack of spiral arms. Like the nuclear bulges of spirals, they are made up mainly of older stars. Because they do not contain much free gas and dust, little star formation takes place within them. Ellipticals can be very large— as much as five times bigger across than typical spirals like our own Galaxy. They can also be very small—dwarf objects made up of only a few hundred thousand stars.

Irregulars and Oddballs

Spiral and elliptical galaxies dominate the universe, but galaxies also come in many other shapes and sizes. Some have an interesting ring structure; others may have had their original structure disrupted by a collision with other galaxies.

Many galaxies have no particular shape at all and are classed as irregulars. They are typically smaller and contain fewer stars than regular spiral and elliptical galaxies. Prime examples of irregulars are the Large and Small Magellanic Clouds prominently visible to the naked eye in the Southern Hemisphere. Recent images of deep-space objects by the Hubble Space Telescope are revealing a whole range of odd-shaped galaxies, which seem to represent evolutionary stages in galaxy formation in the early universe.

The Overactive Galaxies

Most galaxies give off the light, heat, and other radiation we would expect from a concentration of billions of stars. But some give off exceptional energy that cannot easily be accounted for. We call them active galaxies.

Some active galaxies pour out their exceptional energy as light from their centers. They are called Seyferts after U.S. astronomer Carl Seyfert, who first discovered them in the 1940s. Other active galaxies pour out their energy in the form of radio waves, and are known as radio galaxies. One of the most powerful radio sources is the galaxy known as Centaurus A (in the constellation Centaurus), which has a radio output a million times greater than an ordinary galaxy.

Some of the most extraordinary active galaxies are called quasars, for quasi-stellar radio sources. They are known as QSOs, for quasi-stellar objects. They are so called because in the sky they look like stars. But they can't be stars as they can lie billions of light-years away. In reality, they are the bright centers of active galaxies, in the same way that other similar objects named blazars are.

Powering Active Galaxies

Active galaxies pour out so much energy that the usual process of energy production in stars and galaxies— nuclear fusion—cannot possibly explain it. But astronomers know of another power source that can— a supermassive black hole. Like the black hole that results when a big star dies, it comprises a region of space with a powerful gravitational attraction that can swallow up even light. But a supermassive black hole has exceptionally powerful gravity because it contains an extraordinary amount of (invisible) matter—millions of times more than an ordinary black hole. As matter swirls into a supermassive black hole, it heats up and emits the exceptional energy that is the hallmark of Seyferts, quasars, and the other active galaxies.

There also seems to be a supermassive black hole lurking in the center of our own Galaxy. It is pinpointed by a powerful X-ray source known as Sagittarius A. There is evidence that supermassive black holes are present in many other galaxies, though they generally do not produce the exceptional energy of active galaxies.

In the Neighborhood

With the naked eye, we can see only three other galaxies from Earth—the Large and Small Magellanic Clouds and the Andromeda galaxy. The Large Magellanic Cloud is the closest of them at a distance of around 160,000 light-years. The Andromeda galaxy

Centaurus Calling

One of the strangest galaxies in the constellation Centaurus is NGC 5128, which looks as if it is split in two. It is cleft by dark lanes of dense dust. The galaxy pumps out enormous energy as radio waves, and as a radio source it is designated Centaurus A. We now know it is an active galaxy, powered by a supermassive black hole at its center.

Besides our Galaxy and the Andromeda galaxy, there is only one other spiral in the Local Group, the Triangulum galaxy, or M33. The other galaxies are very much smaller, being dwarf elliptical and irregular galaxies.

is the most distant at 2.5 million light-years. Despite such distances, these galaxies are close neighbors in space. With our own Galaxy, they form part of a collection of about 30 galaxies bound loosely together by gravity. We call it the Local Group.

Some of these dwarf galaxies are really tiny. The smallest ones, such as the dwarf elliptical Leo II, measure only about 500 light-years across, 200 times smaller than our own Galaxy. Others, such as the dwarf irregular LGS 3, contain fewer than a million stars, as opposed to around 200 billion in our Galaxy.

Clustering Together

Galaxies group together in a similar way throughout the universe. And small groups in turn form part of much bigger clusters. One of the biggest clusters we know is found in the constellation Coma Berenices. It contains as many as 3,000 galaxies. Over vast expanses of space, these clusters then form part of enormous superclusters. Superclusters are the biggest structures in the universe, measuring hundreds of millions of light-years across.

Our Local Group of galaxies forms part of what is called the Local Supercluster, which is centered on a relatively nearby cluster called the Virgo Cluster. As in many clusters, there are in the center of the Virgo Cluster giant elliptical galaxies that can measure more than 500,000 light-years across. It seems probable that these huge objects grew to such a size by swallowing up smaller galaxies in the cluster.

Close Encounters

The galaxies within clusters tend to be dotted about haphazardly and travel in different directions. From time to time, a galaxy may undergo a close encounter with another or even collide with it. Deep-space images confirm that these events are relatively common.

We can imagine the celestial fireworks that must take place when two galaxies slam into each other at speeds approaching a million miles an hour (1.6 million km/h). The actual stars in the colliding galaxies don't smash into each other like billiard balls because they are too far apart. Instead, they tend to be plucked out of their place in the galaxies by gravitational forces and flung out into space as long streamers. The crashing together of gas clouds within the colliding galaxies provide ideal conditions for star formation, and creates myriad hot new stars.

Which galaxy is affected most in a close encounter or a collision depends on its size and the magnitude of its gravity. For example, if a small galaxy has a close encounter with a large one, it may be torn apart, while the large one will be little affected. When two galaxies of similar size collide, they each lose their original structure and may merge into a single body.

There are some spectacular examples of galactic collisions in the heavens. They include the Antennae

Interacting Spirals

In the constellation Canis Major, two spiral galaxies intertwine as they make a close encounter. Powerful gravitational forces are disrupting the structures of both galaxies and flinging long streamers of stars far into space.

galaxies in Aquarius (page 108) and the Tadpole galaxy in Draco (page 112). The famous Whirlpool galaxy in Canes Venatici is an example of the destruction of a small galaxy by a large one. This was, incidentally, the first galaxy to have its spiral structure identified, by Lord Rosse in Ireland, in 1845.

The Einstein Effect

The combined gravity of galaxy clusters is enormous. And it can have a remarkable effect on how we view very distant astronomical objects. This effect depends on the fact that strong gravity can bend light.

The German-born American physicist Albert Einstein predicted that this would happen in his general theory of relativity, which he put forward in 1915. This came ten years after he had presented his special theory of

Identifying the Galaxies

Galaxies are identified in a similar way to nebulae and clusters (see page 35). They may have a proper name, which can tell us the constellation in which the galaxy is located—for instance the Andromeda galaxy is the most prominent galaxy in the constellation Andromeda. A galaxy may also be named according to what it looks like—the Sombrero galaxy is so named because it looks rather like a Mexican hat. Our own Milky Way Galaxy was named by Greek astronomers, who likened it to a ribbon of milk spurting from the breast of their mythical goddess Hera.

M Some galaxies are also identified by an M number. Messier and his contemporaries saw them as misty patches and thought that they were simply nebulae. The Sombrero galaxy, for example, is M104. It also has an NGC number (4594) because the NGC listing covers galaxies as well as nebulae and clusters.

relativity, which revolutionized our thinking about the universe by linking novel ideas about time, space, motion, mass, and gravitation. In that first theory, he had introduced the most famous of all equations in science, $E=mc^2$ (see page 31).

Gravitational Lensing

The bending of light by the enormous gravity of a galaxy cluster can be compared with the bending of light by a glass lens. We call this phenomenon gravitational lensing.

Acting as a gravitational lens, a galaxy cluster can bring distant objects nearer, just as the lens of a magnifying glass can. The Hubble Space Telescope has returned images of some spectacular gravitational lensing, for example, by the galaxy cluster Abell 2218 (see page 116). Images of distant galaxies, too far away to be detected by telescopes, are brought into focus in the foreground as little arcs.

Usually, astronomers find that the combined gravity of the visible galaxies in a cluster is nowhere near enough to cause the lensing effect. So another source of matter must be present in the cluster. They estimate that it is invisible material they call dark matter, which appears to account for most of the mass in the universe (see page 15).

NGC 4414

Viewed with the naked eye, the constellation Coma Berenices is difficult to make out because even its brightest stars are only of the 4th magnitude. But small telescopes will reveal a host of fuzzy spots among the stars, and larger instruments will resolve them into separate star-island galaxies.

The constellation is home to a huge grouping of galaxies called the Coma Cluster. It contains more than 3,000 galaxies, arranged in a roughly spherical region about 20 million light-years across. It lies about 300 light-years away. But there are also many other galaxies in Coma Berenices that are much closer. One is the galaxy pictured here, NGC 4414.

Like about one-third of the galaxies in the universe, NGC 4414 is a spiral galaxy. It has a bulge of stars in the center, which forms the nucleus. Other stars curve out from the nucleus, forming so-called spiral arms.

Notice that the stars in the central region are predominantly reddish yellow, which means that they are all relatively old. The stars on the spiral arms, on the other hand, are bluer, which means that they are relatively young. This is typical of all spiral galaxies—older stars are found in the nuclear bulge, younger ones in the spiral arms.

Many dark dust lanes are evident in the spiral arms of NGC 4414, silhouetted against the background stars. This shows that there is plenty of material available for future star formation. By comparison, there is precious little star-forming material remaining in the central bulge.

Seen from far away in space, our own galaxy would look similar to NGC 4414, although the spiral arms would not appear so widely separated. In NGC 4414, the spiral arms are well separated, making it an Sc spiral. (Our own Galaxy is an Sb.) With a width of around 56,000 light-years, NGC 4414 is a little over half the size of our own galaxy. It may contain as many as 100 billion stars.

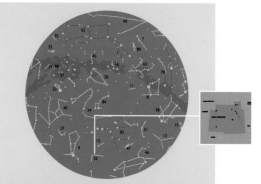

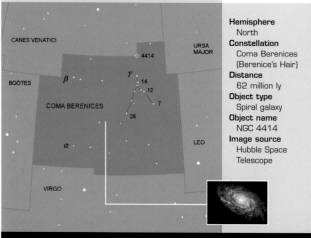

Splendid Spiral

NGC 4414 is a typical spiral galaxy of the Sc type, with quite wide-open spiral arms curving out from the center. It is one of a multitude of galaxies found in Coma Berenices.

Hemisphere
North
Constellation
Coma Berenices
(Berenice's Hair)
Distance
62 million ly
Object type
Spiral galaxy
Object name
NGC 4414
Image source
Hubble Space
Telescope

Spiral arms curve out from the central bulge

Younger stars are found on the spiral arms

This galaxy has more dust than usual in its spiral arms

Bulge of older stars at the center of the galaxy

NGC 4414 85

NGC 1300

Eridanus is sixth largest among the constellations, representing a meandering river. In ancient times it was associated with the River Euphrates in Mesopotamia and the River Nile in Egypt.

Eridanus "rises" near Orion's brilliant star Rigel and flows south to its "mouth" in far southern skies, marked by its brightest star, Achernar. On the way, it makes a broad meander, flowing first to the west, then south, then east again before continuing its journey south. Its passage east is marked by a curve of 4th-magnitude stars identified as Tau (τ) 3, 4, 5, and 6.

Due north of Tau 4, small telescopes can make out the misty patch of the galaxy NGC 1300. Large telescopes show it to be a classic barred-spiral galaxy. This type of galaxy differs from ordinary spirals by having a prominent "bar" of stars running through the central bulge of stars that forms the nucleus. Spiral arms curve out from the ends of the bar.

This superlative 2005 Hubble Space Telescope image shows NGC 1300 in unprecedented detail. It was constructed from exposures taken in September 2004 by Hubble's Advanced Camera for Surveys. Hot blue stars abound in the spiral arms. Bright star clusters are highlighted in red because of the emission of associated glowing hydrogen gas.

Dark dust lanes are evident both in the spiral arms and in the bar through the nucleus. Hubble's sharp eye for detail has pictured a spiral structure within the nucleus. Such a "grand-design" spiral structure had not been seen so clearly before.

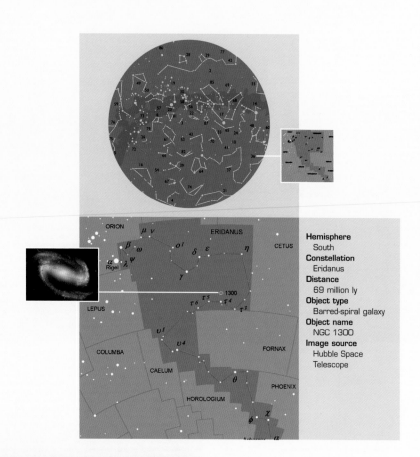

Hemisphere
South
Constellation
Eridanus
Distance
69 million ly
Object type
Barred-spiral galaxy
Object name
NGC 1300
Image source
Hubble Space Telescope

Classic Barred-Spiral

NGC 1300 is regarded as a classic among barred-spiral galaxies. It shows superbly the straight bar of stars through the nucleus and the well-defined spiral arms.

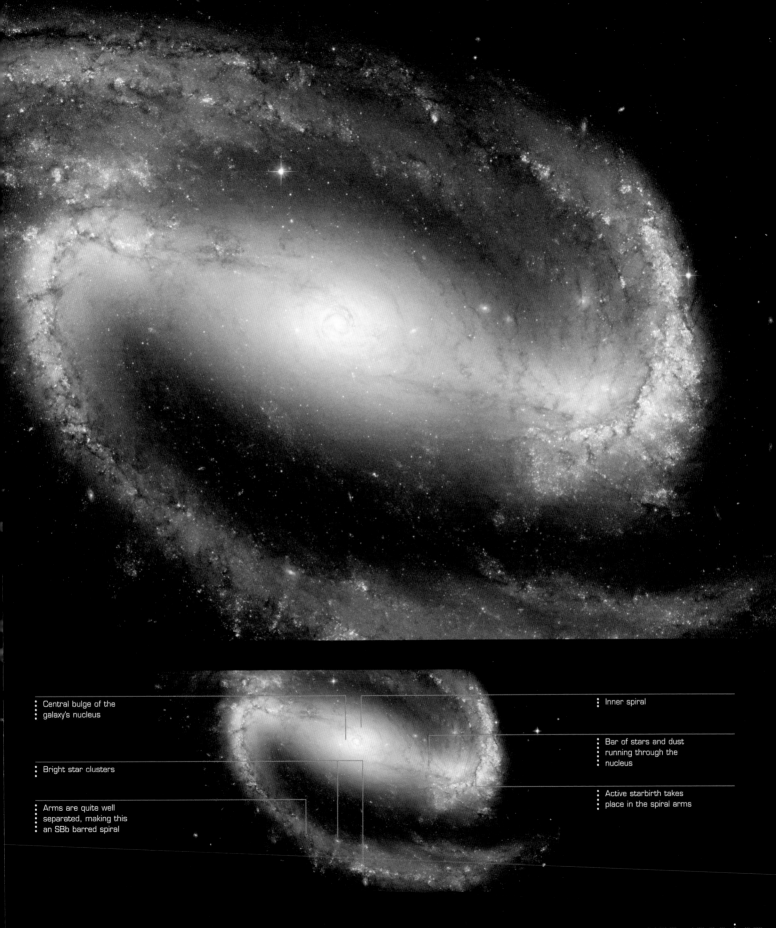

Central bulge of the
galaxy's nucleus

Bright star clusters

Arms are quite well
separated, making this
an SBb barred spiral

Inner spiral

Bar of stars and dust
running through the
nucleus

Active starbirth takes
place in the spiral arms

Sombrero Galaxy : M104

In our telescopes, we can see galaxies galore scattered throughout space. From Earth we see them from different viewpoints, such as head-on (like NGC 1300, pages 86–87) and at an angle (like NGC 4414, pages 84–85).

Here we view a galaxy side view. It is the spiral galaxy M104, also called the Sombrero galaxy because of its resemblance to the type of wide-brimmed hat worn in Mexico. The "brim" is outlined by a dark dust lane running through the middle of the image.

This dark lane represents a slice through the outer disk of the galaxy, which contains the spiral arms. It emphasizes how thin this disk is compared with the nuclear bulge.

We cannot see the spiral arms in this visible-light view because of the obscuring dust lanes. But they can be detected using infrared and radio waves. These studies reveal that the arms curve tightly around the nucleus, making the Sombrero a galaxy of the Sa type.

As spiral galaxies go, M104 is unusual. It has a huge nuclear bulge and an extended fainter region, or halo, around it. The Hubble Space Telescope has discovered nearly 2,000 globular clusters circling around inside the bulge and halo; this is ten times as many as in the bulge of our own Galaxy—but then the Sombrero galaxy is much bigger than our own Galaxy. It is a third larger in diameter and contains billions more stars because of the unusual size of its nuclear bulge.

Top Hat

The well-named Sombrero galaxy can be spotted in small telescopes almost due west of Virgo's brightest star, Spica. But larger instruments are needed to bring out the characteristic dark dust lane across its center.

Fainter "halo" of matter extends far from the nuclear bulge

Central bulge of the galaxy

Dust lane in the disk bisects the galaxy

More distant galaxies shine through the extended halo

The spiral arms of the galaxy are hidden in this view

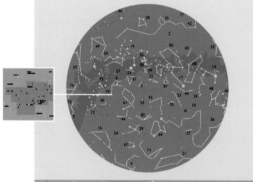

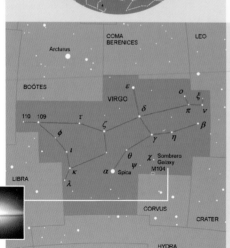

Hemisphere
South
Constellation
Virgo
(the Virgin)
Distance
28 million ly
Object type
Side view of spiral galaxy
Object name
Sombrero galaxy
Image source
Ground observatory (Very Large Telescope, Chile)

Ghost Head Nebula : NGC 2080

Like our own Galaxy, other galaxies are made up of a mix of billions of stars and the billowing clouds of gas and dust in which they are born. They boast stars of all descriptions—stars in their youth blazing brilliantly, stars shining steadily in middle age, and stars in their death throes shrugging off their outer layers or exploding as supernovae and disappearing from the universe in black holes.

In powerful telescopes, we can see in other galaxies the great clouds, or nebulae, of gas and dust that are the birthplace of stars. Our galactic neighbor, the Large Magellanic Cloud, is awash with such nebulae, including the Tarantula, which is visible to the naked eye (see pages 100 to 103).

Just south of the Tarantula is another star-forming region, NGC 2080. This beautiful Hubble image of the area shows how it came to be named the Ghost Head Nebula. Two white eyes appear to be blazing in a human skull.

The white patch that appears to be the ghost's right eye (named A1) is a very hot blob of hydrogen and oxygen lit up by the intense radiation of a single massive star. The ghost's left eye (named A2) is lit up by the radiation from a number of massive stars, embedded in dust.

Astronomers estimate that the massive stars in A1 and A2 must have formed within the last 10,000 years. They are still shrouded in the gas clouds of their birth, and their powerful stellar winds have not had time to blow the gas away.

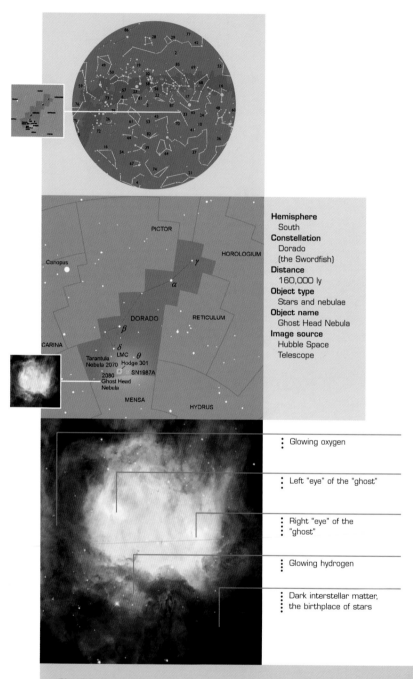

Hemisphere
South
Constellation
Dorado
(the Swordfish)
Distance
160,000 ly
Object type
Stars and nebulae
Object name
Ghost Head Nebula
Image source
Hubble Space
Telescope

Glowing oxygen

Left "eye" of the "ghost"

Right "eye" of the "ghost"

Glowing hydrogen

Dark interstellar matter, the birthplace of stars

Ghoulish Apparition

Little imagination is required to see how this nebula got its name. The Hubble Space Telescope has picked out amazing detail in this vigorous star-forming region in our neighboring galaxy.

Ghost Head Nebula : 91

Lindsay–Shapley Ring Galaxy

The majority of galaxies we find in the universe are either spirals or ellipticals. But we also find many that have other kinds of structures. Among the most interesting are the ring galaxies. An example is the Lindsay–Shapley Ring galaxy (AM 0644-741) shown here.

Ring galaxies are so called because their nucleus is located inside, and essentially separated from, a ring of stars. In this particular ring galaxy, the nucleus is way off-center. Astronomers estimate that most ring galaxies form as a result of collisions between galaxies.

Originally, the galaxy that we see now as AM 0644-741 was an ordinary spiral, with a bulging nucleus and a surrounding disk containing the spiral arms. But several billion years ago, another galaxy (the "intruder") plunged through the disk of the spiral (the "target").

The shock waves set up by the collision totally disrupted the target galaxy's structure. It sent the stars and gas clouds in the spiral arms rushing outward, forming an ever-expanding ring. As the ring plowed into the surrounding interstellar medium, the gas got compressed and then collapsed under its own gravity to form an abundance of new stars.

The outburst of new star formation explains why the ring is blue. It is full of young massive stars that are very hot and emit blue–white light. This contrasts with the stars in the nucleus, which are much older and yellower.

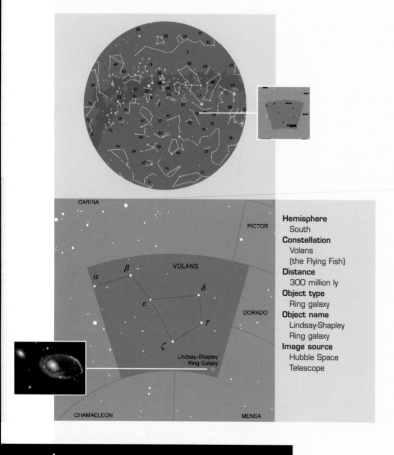

CARINA

PICTOR

VOLANS

β

α

δ

ε

DORADO

γ

ζ

Lindsay-Shapley
Ring Galaxy

CHAMAELEON

MENSA

Hemisphere
South
Constellation
Volans
(the Flying Fish)
Distance
300 million ly
Object type
Ring galaxy
Object name
Lindsay-Shapley
Ring galaxy
Image source
Hubble Space
Telescope

Sparkling Ring

The ring of hot stars in the Lindsay–Shapley Ring Galaxy measures some 150,000 light-years in diameter. This makes it half as big again as our own Galaxy.

More distant large
spiral galaxy

Ring around nucleus

Distant galaxy

Nearby stars in
our own Galaxy

Hot blue stars

Nucleus of original
galaxy

Yellow older stars

Hoag's Object

This beautiful head-on galaxy is known as Hoag's Object. It is another example of a ring galaxy. U.S. astronomer Art Hoag discovered the object in 1950 and thought at first that it was a planetary nebula. A planetary nebula forms when a dying Sunlike star puffs off its outer layers, often producing a ringlike structure. But it was soon realized that the object must be a galaxy.

Hoag's Object measures about 120,000 light-years across, somewhat larger than our own Galaxy. A broad ring of gas and young hot blue stars surrounds the nucleus of older, more yellow stars. The whole structure is remarkably symmetrical, unlike that of the Lindsay–Shapley Ring galaxy on the previous pages.

Most ring galaxies form when an "intruder" galaxy collides with another "target" galaxy and disrupts its structure. Stellar material from the target forms a ring that expands away from the nucleus. Within the ring, new generations of stars are born, shining blue–white in their infancy.

However, in the case of Hoag's Object there seems to be no trace in the surrounding space of an intruder galaxy. Astronomers have suggested that the ring itself may be the shredded remains of the intruder.

In Hoag's Object and most other ring galaxies, the ring of stars forms in the same plane as the nucleus (and disk) of the original target galaxy. But, in a few galaxies, the ring forms in a plane perpendicular to the plane of the nucleus—in other words, over the poles of the nucleus. These are known as polar ring galaxies. No one really knows how such structures form.

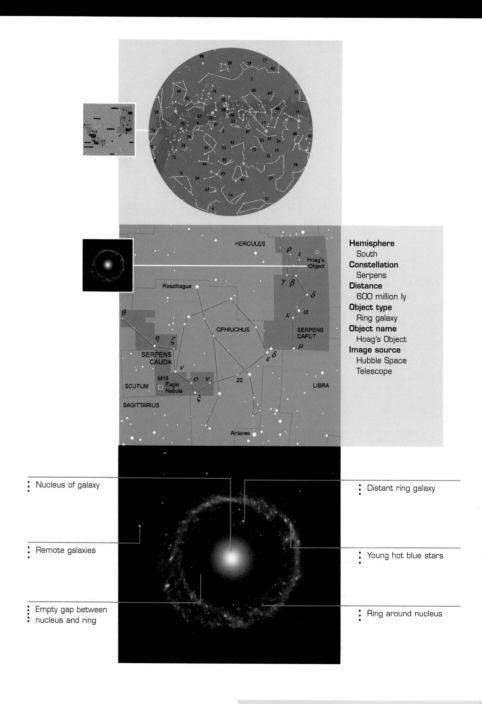

Hemisphere
South
Constellation
Serpens
Distance
600 million ly
Object type
Ring galaxy
Object name
Hoag's Object
Image source
Hubble Space Telescope

Nucleus of galaxy

Remote galaxies

Empty gap between nucleus and ring

Distant ring galaxy

Young hot blue stars

Ring around nucleus

Wheel within a Wheel

What appears to be another ring galaxy is visible in the gap between the nucleus and the stellar ring of Hoag's Object. It is one of several background galaxies in the picture.

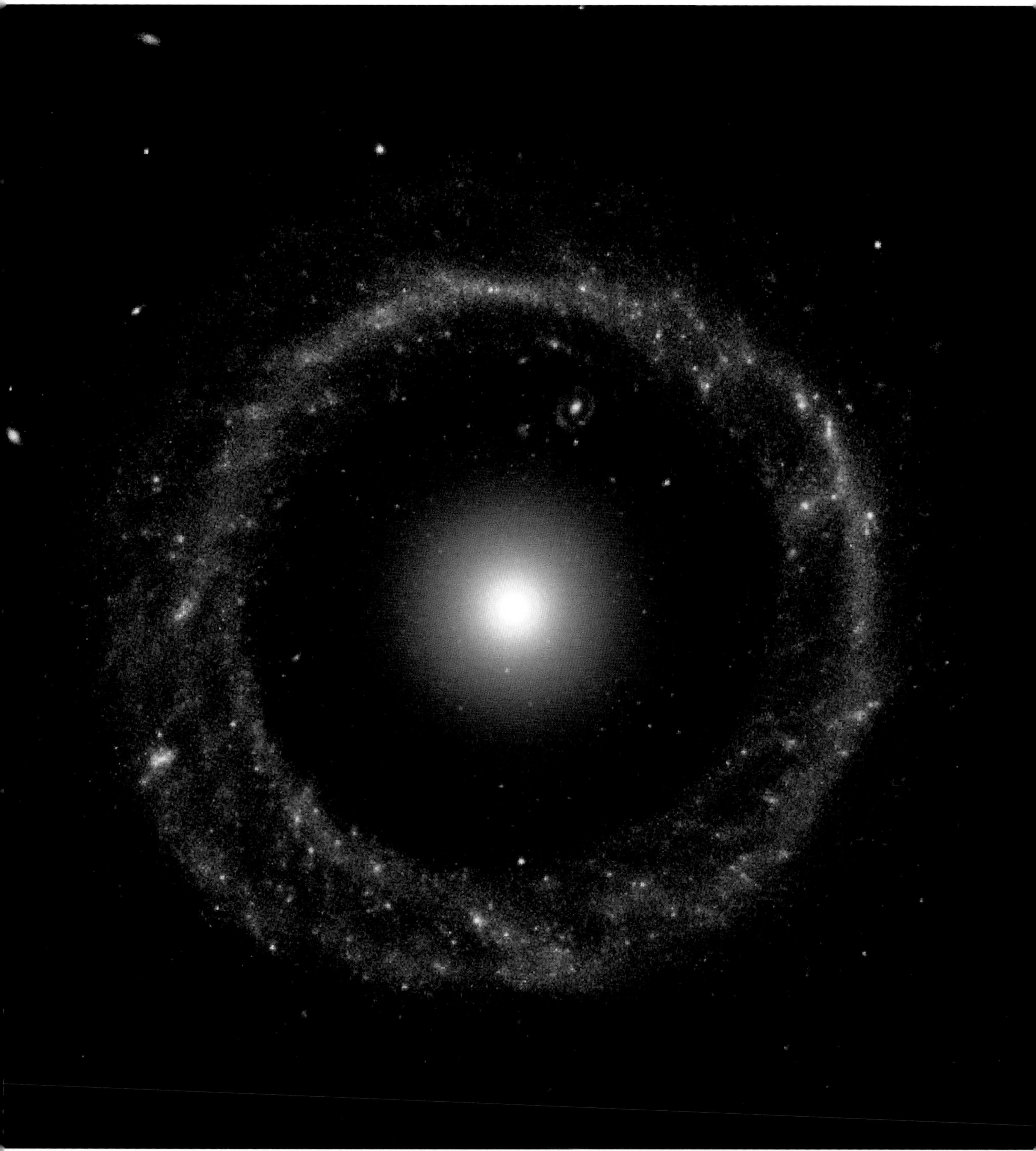

NGC 7742

Our own Galaxy is a typical spiral galaxy, similar to many other spirals in the universe. As spirals go, it is relatively large, measuring some 100,000 light-years across. It is as bright as would be expected from its 200 billion stars.

The galaxy NGC 7742, pictured here by the Hubble Space Telescope, is a much smaller spiral galaxy, only about a third the size of our own Galaxy. It has a different structure from our own Galaxy, but more important, it has an exceptionally bright core, or nucleus. The bright nucleus pours out as much light as all the stars in our own Galaxy.

Astronomers know of many galaxies similar to NGC 7742. They are known as Seyfert galaxies, named for U.S. astronomer Carl Seyfert. They are one type of a class of objects known as active galaxies. Others include radio galaxies and quasars.

Study of the energy emitted by the nucleus of Seyfert galaxies indicates that it comes from a power source only about one light-year across and that the energy is emitted by a disk of intensely hot gas, whirling around at high speed.

Astronomers think that the hot gas is whirling around as it spirals into a black hole. This is not the "ordinary" kind of black hole produced when a big and heavy star dies (see page 35). Rather it is a supermassive black hole produced when material in a gas cloud collapses, typically at the center of a galaxy.

Whereas an "ordinary" black hole results from the collapse of the core of a star with the mass of just a few suns, a supermassive black hole results from the collapse of material with the mass of between a million and a billion suns.

Sunnyside Up

Appearing like a sunnyside-up egg on a celestial plate is this beautiful head-on small spiral galaxy, NGC 7742. Its brilliant center confirms that it is an active galaxy of the Seyfert type.

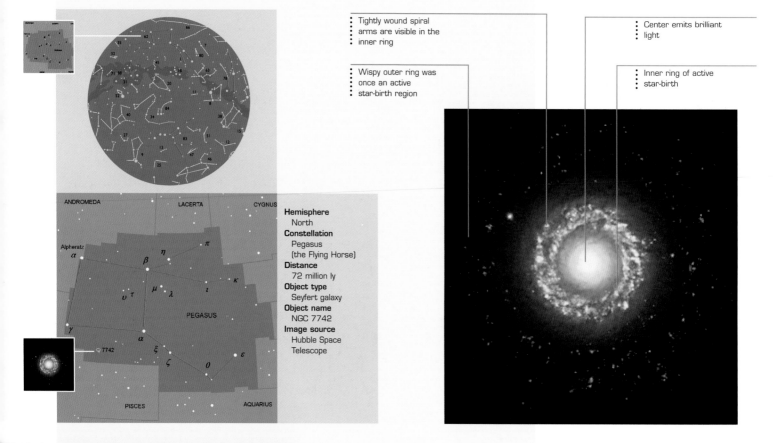

Tightly wound spiral arms are visible in the inner ring

Wispy outer ring was once an active star-birth region

Center emits brilliant light

Inner ring of active star-birth

Hemisphere
North
Constellation
Pegasus
(the Flying Horse)
Distance
72 million ly
Object type
Seyfert galaxy
Object name
NGC 7742
Image source
Hubble Space
Telescope

ANDROMEDA
LACERTA
CYGNUS
Alpheratz
α
η
π
β
κ
υ τ
μ
λ
ι
PEGASUS
γ
α
ξ
ε
7742
ζ
θ
PISCES
AQUARIUS

NGC 7742 97

Milky Way Galaxy

The center of our Galaxy—the Milky Way Galaxy—lies in the direction of the constellation Sagittarius in far southern skies. We can't look at or photograph the central region in ordinary visible light because clouds of dust obscure our view. But we can "see" into the center if we use telescopes that are sensitive to radio waves or infrared light. These waves can penetrate dust clouds.

This image of the center of our Galaxy at infrared wavelengths was returned by the Very Large Telescope in Chile. It shows a cluster of hot and cool stars and dust clouds between them. It covers a region about 2.5 light-years across.

The arrows in the middle pinpoint a powerful radio and X-ray source known as Sagittarius A* (SgrA*). Astronomers believe that this marks the location of a black hole. It is a supermassive black hole containing the mass of some 3 billion suns (see page 96).

Measurements show the stars close to SgrA* circle around it at amazing speeds—as high as 3,000 miles (5,000 km) a second. This is what would be expected if they were accelerated by the intense gravity around a black hole.

Radio images of the center of the Galaxy reveal many interesting structures. For example, there is a ring of magnetized gas, called the radio lobe, a few hundred light-years across. And there is a circle of huge molecular clouds known as the molecular ring about 1,000 light-years across, in which intense star formation is taking place.

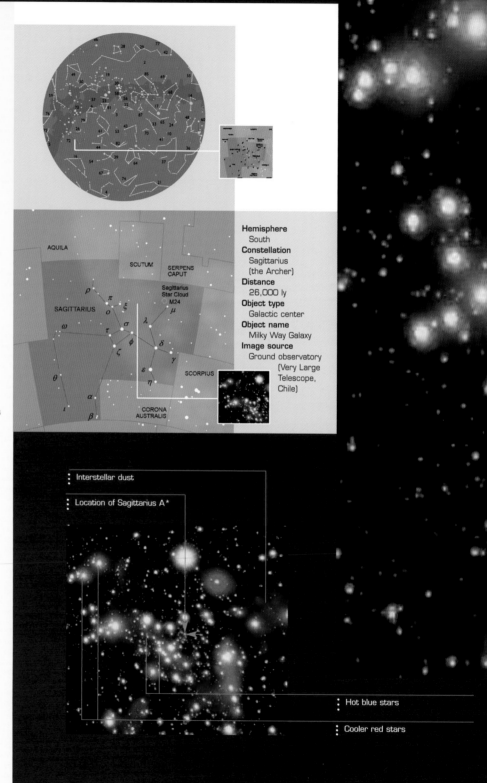

Hemisphere
South
Constellation
Sagittarius
(the Archer)
Distance
26,000 ly
Object type
Galactic center
Object name
Milky Way Galaxy
Image source
Ground observatory
(Very Large
Telescope,
Chile)

Interstellar dust

Location of Sagittarius A*

Hot blue stars

Cooler red stars

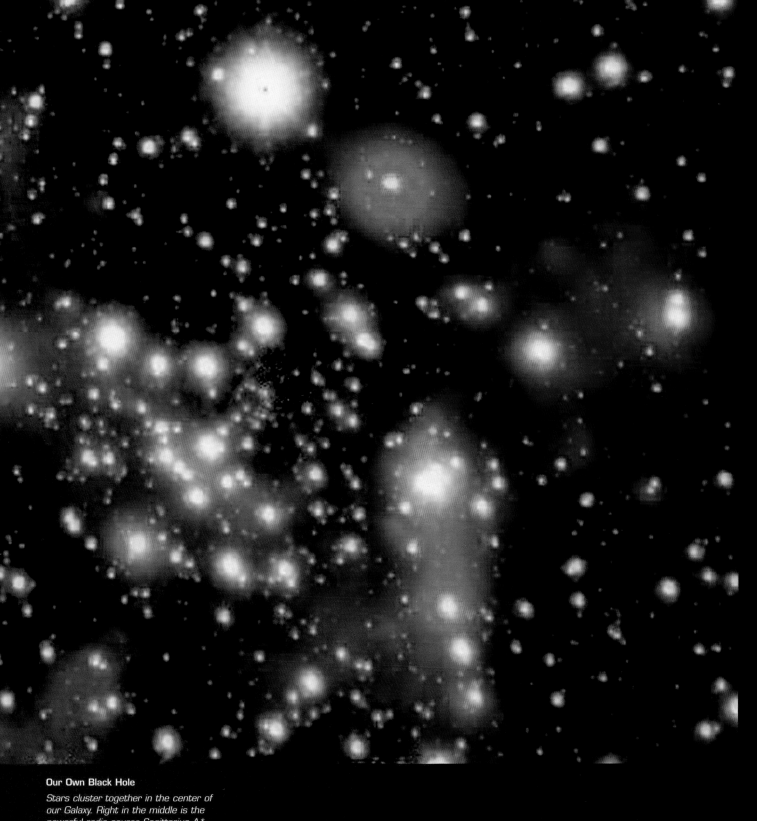

Our Own Black Hole

*Stars cluster together in the center of
our Galaxy. Right in the middle is the
powerful radio source Sagittarius A*,
marking the location of a black hole.*

In far southern skies, we can see two misty patches with the naked eye, which look rather like wispy clouds in the atmosphere. But these clouds lie far beyond Earth. They are separate galaxies, known as the Large Magellanic Cloud (LMC) and the Small Magellanic Cloud (SMC). They are named for the Portuguese navigator Ferdinand Magellan, who was the first European to record the Clouds when he sailed the southern seas in the early 1500s.

Both Clouds are in orbit around our own Galaxy—we call them satellite galaxies. Over time, they will merge with our own Galaxy.

The LMC and the SMC are classed as irregular galaxies because they lack any definite structure. At a distance of about 160,000 light-years, the LMC is closest to us—the SMC lies about 30,000 light-years farther away. The highlight of the LMC is one of the largest star-forming regions we know. We call it the Tarantula Nebula, also known as 30 Doradus.

The Tarantula, which measures about 800 light-years across, boasts glowing filaments that vaguely resemble the legs of a spider. This is how it got its name. It is a typical emission nebula, in which hydrogen gas is set glowing by the intense ultraviolet radiation of embedded stars.

The stars responsible for the light emission in the Tarantula Nebula are young, hot, and massive. They form a cluster known as R136. Other smaller clusters are found throughout the nebula, lighting up the surrounding gas.

NGC 2100 is a giant cluster of blue–white stars only about 20 million years old. They contrast with the stars in the globular cluster KMHK 1137, which are redder, fainter, and older. ➤

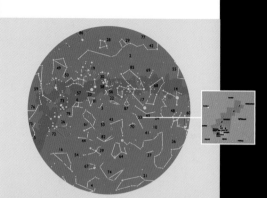

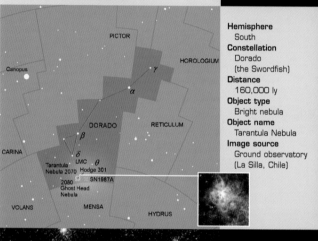

Tantalizing Tarantula
Wide-field view of the Tarantula Nebula, photographed from the European Southern Observatory's La Silla Observatory in Chile. It clearly shows the spidery shape of this massive star-forming region, which is visible to the naked eye.

Hemisphere
South
Constellation
Dorado
(the Swordfish)
Distance
160,000 ly
Object type
Bright nebula
Object name
Tarantula Nebula
Image source
Ground observatory
(La Silla, Chile)

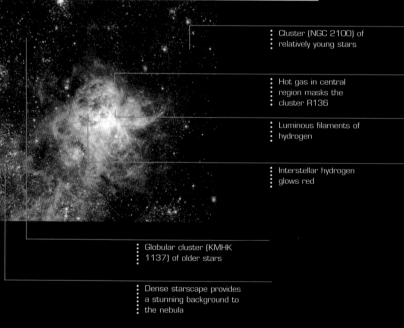

Cluster (NGC 2100) of relatively young stars

Hot gas in central region masks the cluster R136

Luminous filaments of hydrogen

Interstellar hydrogen glows red

Globular cluster (KMHK 1137) of older stars

Dense starscape provides a stunning background to the nebula

➤ This spectacular Hubble Space Telescope picture of the Tarantula Nebula is a mosaic of many images. It is centered on the star cluster R136. This is a relatively young cluster, born about two million years ago.

R136 is a cluster of several dozen massive stars, each with a mass up to 100 times that of the Sun. These stars are very hot, with a surface temperature of 100,000°F (55,000°C)—ten times that of the Sun.

Being so young and so hot, the stars in R136 give off powerful stellar winds and intense radiation. This sets the gas in the surrounding region into turmoil and triggers it into emitting light and other radiation. The turbulent gases form a delicate tracery of glowing filaments.

Dense regions of interstellar material form dark columns, or pillars, here and there. Stars are forming within these so-called "elephant trunks," and will eventually be exposed when the surrounding gas is evaporated by the stellar winds and intense radiation of massive stars behind them.

Cauldron of Creation

In this region of the Tarantula Nebula in the Large Magellanic Cloud, stars are being born in their millions. The interstellar gas is set glowing by the radiation emitted by hot, newborn stars.

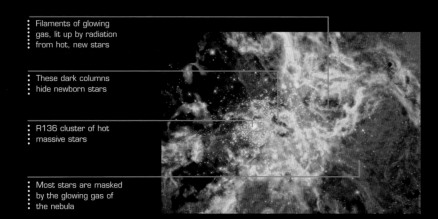

Filaments of glowing gas, lit up by radiation from hot, new stars

These dark columns hide newborn stars

R136 cluster of hot massive stars

Most stars are masked by the glowing gas of the nebula

Andromeda Galaxy : M31

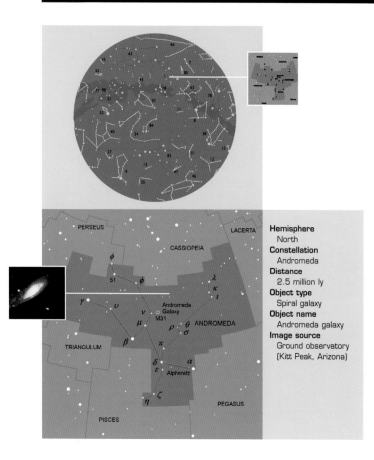

Hemisphere
North
Constellation
Andromeda
Distance
2.5 million ly
Object type
Spiral galaxy
Object name
Andromeda galaxy
Image source
Ground observatory
(Kitt Peak, Arizona)

Apart from the two Magellanic Clouds, only one other galaxy is visible to the naked eye. We can see it as a faint misty patch in the northern constellation Andromeda, M31. It lies very much farther away than the Magellanic Clouds, at a distance of some 2.5 million light-years.

The reason we can see the Andromeda galaxy so far away is that it is very large. It measures about 150,000 light-years across (compared with 100,000 light-years for our own Galaxy) and is made up of as many as 400 billion stars.

The Andromeda galaxy is a spiral galaxy like our Galaxy, with curved arms forming out of a central nucleus. We can't see the arms very well from Earth because we look at the galaxy nearly from the side.

Located close to the galaxy are two satellite galaxies—ones that are in orbit around it. The closer is M32, the more distant NGC 205. Both are dwarf elliptical galaxies. M32 measures only about 5,000 light-years across; NGC 205 is roughly twice as big.

Historically, the Andromeda galaxy is very important. It was once considered to be a spiral nebula within our own Galaxy. But, in 1923, U.S. astronomer Edwin Hubble proved that this "nebula" lay far beyond the confines of our own Galaxy and had to be a galaxy in its own right. He went on to prove that other spiral nebulae were far-distant galaxies, or "extragalactic nebulae," as he preferred to call them.

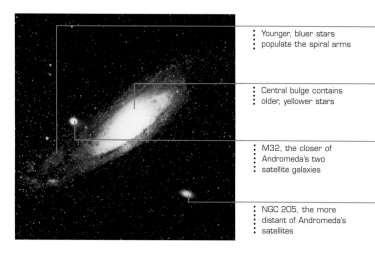

Younger, bluer stars populate the spiral arms

Central bulge contains older, yellower stars

M32, the closer of Andromeda's two satellite galaxies

NGC 205, the more distant of Andromeda's satellites

Largest in the Local

The Andromeda galaxy, which we see nearly from the side from Earth. This makes it difficult to make out its spiral arms. The largest galaxy in the Local Group, Andromeda is half as wide again as our own Galaxy.

NGC 604

Our own Galaxy and the three outer galaxies we can see in the heavens with the naked eye (the two Magallanic Clouds and the Andromeda galaxy) are loosely bound together by gravity and form part of a small cluster of about 30 galaxies called the Local Group.

The Andromeda galaxy is the biggest member of the Group, followed by our own Galaxy. Both are spiral galaxies. The next largest member of the group, M33 in the constellation Triangulum, is also a spiral. The other members of the group are much smaller and are elliptical or irregular in shape.

M33, also called the Triangulum galaxy, is another favorite among astronomers because we see it head-on and have a good view of its wide-open spiral arms. Visually, it has a magnitude of just over six, which puts it just outside naked-eye visibility but well within the reach of binoculars and small telescopes.

A striking feature of M33 is a bright nebula in one of its spiral arms. Designated NGC 604, it is a huge star-forming region. Measuring some 1,300 light-years across, it is a hundred times the size of the Orion Nebula in our own Galaxy. And, whereas the Orion Nebula is lit up primarily by the radiation from the four brightest stars of the Trapezium, NGC 604 is lit up by more than 200 bright stars.

Most of these stars lie in the heart of NGC 604. Hot and blue, they are up to 120 times more massive than the Sun, with a surface temperature up to 72,000°F (40,000°C). They are only about 3 million years old. Their powerful stellar winds have carved out a huge cavity within the nebula, and their intense ultraviolet radiation have made the nebula gases glow.

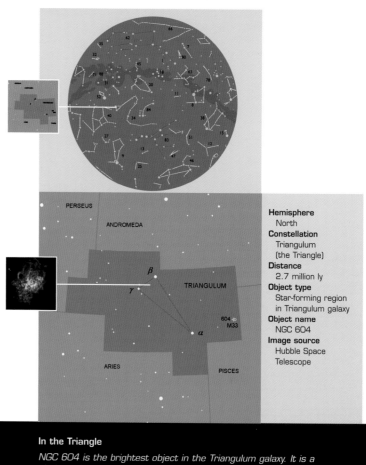

Hemisphere
North
Constellation
Triangulum
(the Triangle)
Distance
2.7 million ly
Object type
Star-forming region
in Triangulum galaxy
Object name
NGC 604
Image source
Hubble Space
Telescope

In the Triangle

NGC 604 is the brightest object in the Triangulum galaxy. It is a glorious object when viewed in the most powerful telescopes, as here by the Hubble Space Telescope. Hot stars light up the hydrogen.

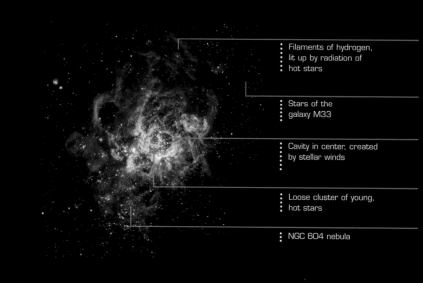

Filaments of hydrogen, lit up by radiation of hot stars

Stars of the galaxy M33

Cavity in center, created by stellar winds

Loose cluster of young, hot stars

NGC 604 nebula

NGC 604 | 107

Antennae Galaxies

Galaxies are not scattered haphazardly throughout the universe. Rather, they are found in groups or clusters. But within these clusters, they pursue relatively independent paths. From time to time, two galaxies may pass close to each other or even collide. When this happens, a brilliant celestial fireworks show takes place.

In the constellation Corvus, we see the result of a collision between two galaxies that took place about a billion years ago. The two galaxies, NGC 4038 and 4039, are still intertwined, and have two long curved tails of stars and dust streaming out from them.

The colliding galaxies are called the Antennae because the curved tails look like an insect's antennae. This is clear from the ground-based observatory view of the galaxies, shown on the left of the picture.

In the main image, the Hubble Space Telescope has revealed the fireworks triggered by the colliding galaxies. Remnants of the bright nuclei of the two galaxies are still evident, but little remains of the structure of the galaxies, which were probably originally ordinary spirals.

The predominantly blue–white color of the stars in the image show that they are very hot and very young. In many places, hundreds of thousands of stars have been born together and form dense clusters. The Hubble Space Telescope has identified over 1,000 such clusters, all born when the gas clouds in the two galaxies crashed into one another.

Collisions between galaxies seemed to be much more common in the early days of the universe. This is borne out by the Hubble Deep Field image (pages 118–119), which reveals a host of odd-shaped galactic objects that seem to be collided galaxies.

Interstellar Fireworks

The Antennae galaxies, seen in an image taken by a conventional telescope (left) and a stunning Hubble Space Telescope image of the central region. The conventional image shows the long tails of luminous matter that give the object its name.

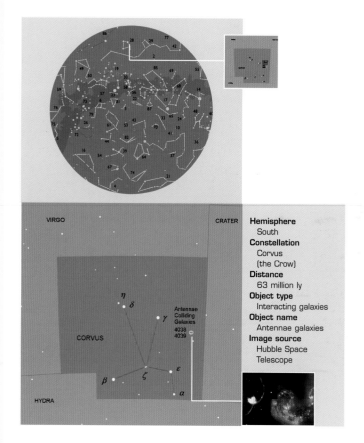

Hemisphere
South
Constellation
Corvus
(the Crow)
Distance
63 million ly
Object type
Interacting galaxies
Object name
Antennae galaxies
Image source
Hubble Space Telescope

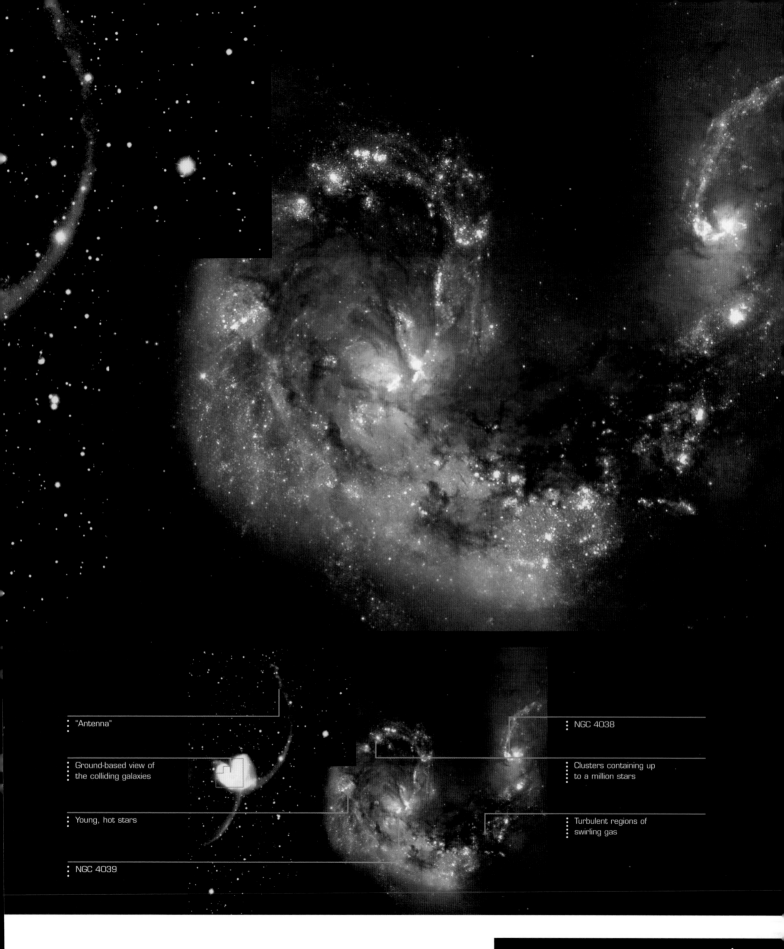

"Antenna"

Ground-based view of
the colliding galaxies

Young, hot stars

NGC 4039

NGC 4038

Clusters containing up
to a million stars

Turbulent regions of
swirling gas

Whirlpool Galaxy : M51

With a visual magnitude of around 8, Messier object number 51 can be spotted in binoculars and small telescopes. It is easy to find because it lies just south of the first star in the handle of the Big Dipper (Plow) in Ursa Major.

Larger telescopes show that M51 is a glorious head-on spiral galaxy with wide-open spiral arms. It is well named the Whirlpool galaxy.

This picture of M51 is in false color. It is a combination of a radio image taken by the Very Large Array radio telescope in New Mexico and a visible-light image taken by a conventional optical telescope. Combining images at different wavelengths enables astronomers to extract much more detail from a subject.

It is clear that the Whirlpool is no ordinary galaxy. Compared with ordinary spirals, it is distorted and in addition has a bright object located at the end of one of its arms. This object is the remains of another smaller galaxy. So the Whirlpool is actually two galaxies in one. The main one is designated NGC 5194, the smaller one NGC 5195.

About 300 million years ago, NGC 5195, an ordinary spiral, strayed into the path of the much larger NGC 5194 and collided with it. The powerful gravity of the large galaxy tore the small one apart, and most of its stellar matter formed a bridge with the large one that we see today. At the same time, the shock of the galactic collision triggered off a crescendo of star formation in NGC 5194 and distorted its structure.

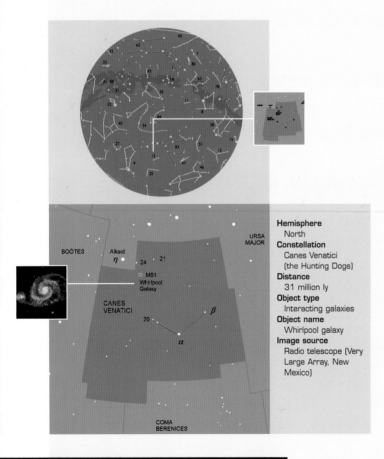

Hemisphere
North
Constellation
Canes Venatici
(the Hunting Dogs)
Distance
31 million ly
Object type
Interacting galaxies
Object name
Whirlpool galaxy
Image source
Radio telescope (Very Large Array, New Mexico)

In the Whirlpool

The aptly named Whirlpool galaxy, shown in false color. It was once thought to be a nebula in our own Galaxy. Irish astronomer Lord Rosse discovered it had a spiral structure in 1845, decades before astronomers realized that spiral "nebulae" were in fact separate galaxies.

Spiral galaxy NGC 5194

Spiral galaxy NGC 5195

Green shows stars

Blue shows gas

Red shows regions of strong magnetism

Tadpole Galaxy

Thousands of distant galaxies form the dramatic backdrop for this peculiar galaxy, named the Tadpole because of its shape. A distorted spiral galaxy forms the Tadpole's head, and a long streamer of stars its tail.

The Tadpole was once an ordinary spiral galaxy, but, eons ago, a small galaxy collided with it. Powerful gravitational interaction between the two galaxies distorted the large galaxy's structure and created the tail of stellar debris, which stretches for nearly 300,000 light-years.

Clusters of blue stars, spawned in the galaxy collision, are visible both in the spiral arms of the large galaxy and in the debris tail. These clusters may contain as many as a million massive stars that are ten times hotter and a million times brighter than our Sun. These clusters will redden with age and will eventually become globular clusters orbiting the center of the Tadpole, similar to the ones we find orbiting the center of our own Galaxy.

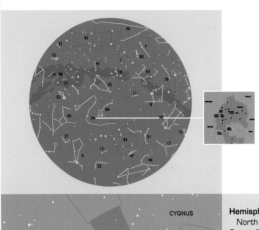

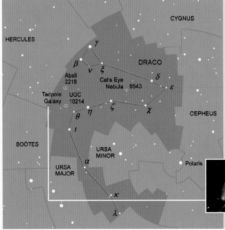

Hemisphere
North
Constellation
Draco (the Dragon)
Distance
420 million ly
Object type
Peculiar galaxy
Object name
Tadpole galaxy
Image source
Hubble Space
Telescope

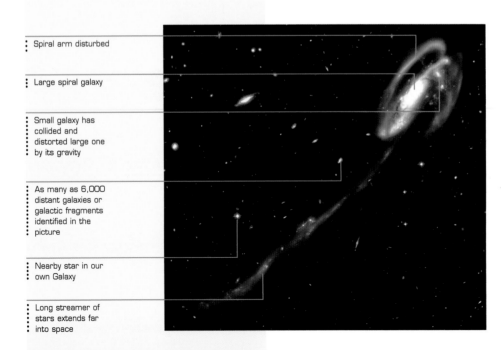

Spiral arm disturbed

Large spiral galaxy

Small galaxy has collided and distorted large one by its gravity

As many as 6,000 distant galaxies or galactic fragments identified in the picture

Nearby star in our own Galaxy

Long streamer of stars extends far into space

Galactic Tadpole

This distorted spiral galaxy, well named the Tadpole, is one of the most spectacular Hubble Space Telescope images. It was one of the first pictures taken by the telescope's Advanced Camera for Surveys, installed in 2002.

Hickson Compact Group 87

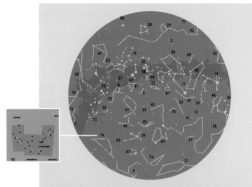

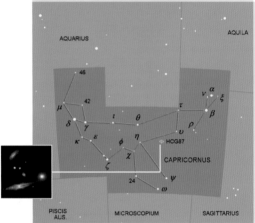

On a large scale, galaxies gather together into often very big clusters, bound loosely together by gravity. But within these clusters there exist much smaller groups of galaxies, bound together more tightly by their mutual gravity. An example is Hickson Compact Group 87, pictured here by the Hubble Space Telescope.

The group comprises four galaxies, three of which are close together. The largest galaxy (87a) is a distorted spiral galaxy that we see from the side. It has prominent dust lanes that run along the plane of the disk that contains the spiral arms.

Close by is a small elliptical galaxy (87b), almost spherical in shape. The large spiral seems to be drawing stellar material from the elliptical galaxy, forming a bridge between the two.

Both these two galaxies have exceptionally bright centers, or nuclei. Astronomers estimate that they have active galactic nuclei—nuclei in which supermassive black holes are pouring out extraordinary energy (see page 96).

The other two galaxies in the group, a small one (87c) located above the spiral seen from the side and a larger one (87d) farther out, are also spirals. We see the small one head-on, with well-separated spiral arms. It seems to be undergoing a burst of star formation, no doubt triggered by the gravitational disturbance of its close neighbors. We see the 87d spiral at more of an angle, but can still make out its spiral arms and its compact, bright nucleus.

Mutual gravitational attraction between the four galaxies not only triggers new star formation within them and causes structural distortion, but it will also eventually bring the galaxies together to form a single large elliptical galaxy.

Hemisphere
South
Constellation
Capricornus
(the Sea Goat)
Distance
400 million ly
Object type
Galaxy cluster
Object name
Hickson Compact
Group 87
Image source
Hubble Space
Telescope

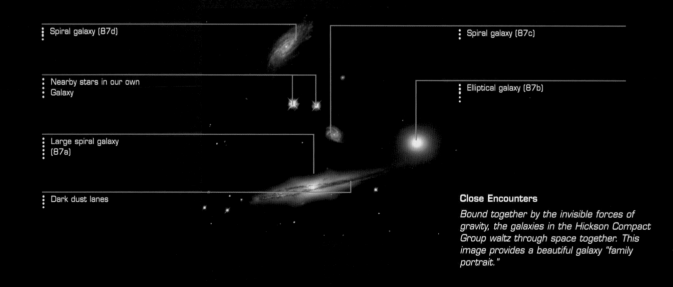

Spiral galaxy (87d)

Nearby stars in our own Galaxy

Large spiral galaxy (87a)

Dark dust lanes

Spiral galaxy (87c)

Elliptical galaxy (87b)

Close Encounters

Bound together by the invisible forces of gravity, the galaxies in the Hickson Compact Group waltz through space together. This image provides a beautiful galaxy "family portrait."

Abell 2218

This huge cluster of as many as 10,000 galaxies in Draco provides an outstanding example of the phenomenon astronomers call gravitational lensing. It acts in effect like a gigantic cosmic magnifying glass, bringing into view objects too faint and too remote to be seen directly in telescopes.

The explanation for this magnifying effect is gravity, or rather the immensely powerful combined gravity of the thousands of galaxies in the Abell 2218 cluster. This enormous gravity has the ability to bend light rays, as predicted by Albert Einstein in his General Theory of Relativity (1916).

Gravitational lensing in Abell 2218 throws up many such arcs, showing remote galaxies at distances up to more than 10 billion light-years. This reveals them as they were when the universe was only a few billion years old. Some are protogalaxies—star systems that have not yet organized themselves into regular galaxies.

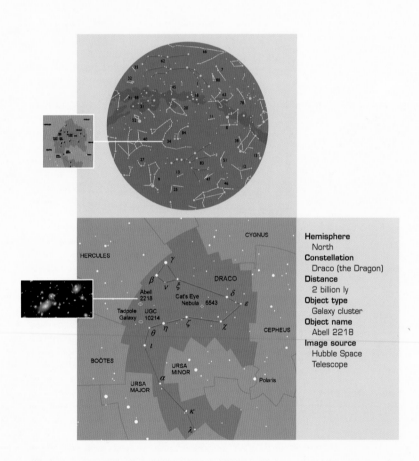

Hemisphere	
North	
Constellation	
Draco (the Dragon)	
Distance	
2 billion ly	
Object type	
Galaxy cluster	
Object name	
Abell 2218	
Image source	
Hubble Space Telescope	

Galaxy Clusters

The combined mass of this cluster of galaxies in Draco, called Abell 2218, is enormous. Because of this mass, the cluster has collectively a very powerful gravity. So much so that this gravity can bend the light from even more remote galaxies and bring it into focus in the image. Lensing produces images of remote galaxies as a series of little circular arcs, like those in the picture. They are distorted into arcs by the lensing process.

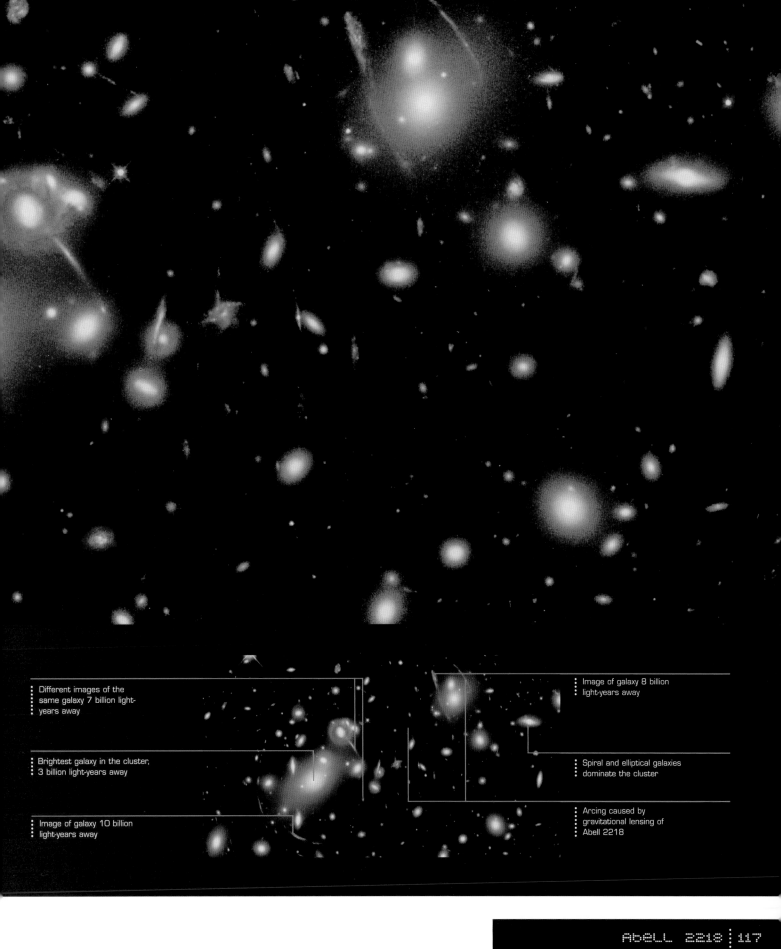

Different images of the
same galaxy 7 billion light-
years away

Brightest galaxy in the cluster,
3 billion light-years away

Image of galaxy 10 billion
light-years away

Image of galaxy 8 billion
light-years away

Spiral and elliptical galaxies
dominate the cluster

Arcing caused by
gravitational lensing of
Abell 2218

Hubble Deep Field

On December 8, 1995, the Hubble Space Telescope locked onto a tiny region of space just north of the star in the Big Dipper we call Delta (δ) Ursae Majoris, or Megrez. From Earth, this region appears only as big as a pinhead. Over the next 20 days, the Telescope made 342 separate exposures—totaling more than 120 hours—studying this pinhead of space.

The result was a ground-breaking image known as the Hubble Deep Field. It revealed a bewildering assortment of more than 1,500 galaxies never seen before because they are so faint—billions of times fainter than can be seen with the naked eye. The Telescope's accumulated long exposure time had allowed the faint light from these galaxies to build up until they became visible.

The galaxies viewed in the Hubble Deep Field are so faint because they are so far away. The most distant appear to be up to 11 billion light-years away. In other words, we are seeing them as they were just a few billion years after the universe began. Studying the most distant and nearer galaxies allows astronomers to gain an idea of how galaxies evolve.

The image shown here is the result of a revisit of the Hubble Deep Field region, produced from exposures made between December 1997 and June 1998. With an accumulated exposure time of about six days, the image is virtually identical with the original one produced two years earlier.

But there is a notable exception, highlighted in a box here. In the center is a faint red elliptical galaxy. When Hubble scientists compared the light output from this galaxy between 1995 and 1997, they discovered a bright blob in the 1997 image that was caused by a supernova. They estimate that it took place 10 billion years ago.

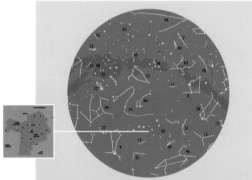

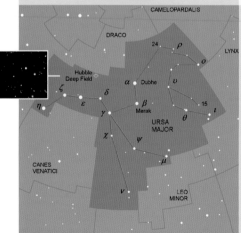

CAMELOPARDALIS

DRACO

LYNX

Hubble Deep Field

Dubhe

URSA MAJOR

Merak

CANES VENATICI

LEO MINOR

Hemisphere
North
Constellation
Ursa Major
(the Great Bear)
Distance
10 billion ly
Object type
Galaxy field
Object name
Hubble Deep Field
Image source
Hubble Space
Telescope

Blast from the Past

In this 1997 image, the Hubble Space Telescope revisits the Hubble Deep Field region studied two years earlier. It spies an object not present before, a supernova blast from the early universe.

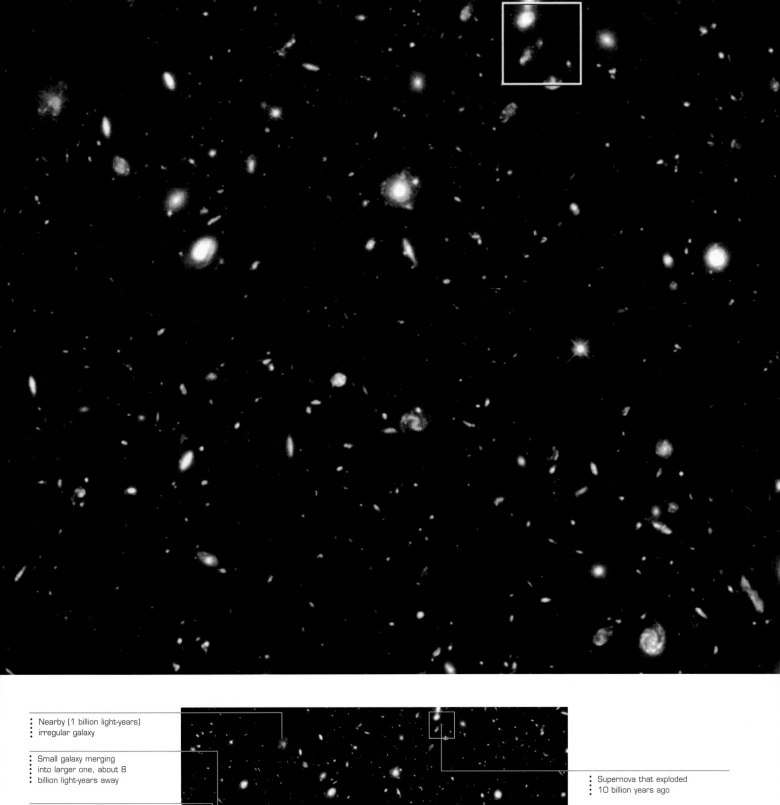

Nearby (1 billion light-years)
irregular galaxy

Small galaxy merging
into larger one, about 8
billion light-years away

Nearby star in our own
Galaxy

Spiral galaxy about 6
billion light-years away

Supernova that exploded
10 billion years ago

Developing spiral galaxy

Galaxy in early stages of
evolution, over 10 billion
light-years away

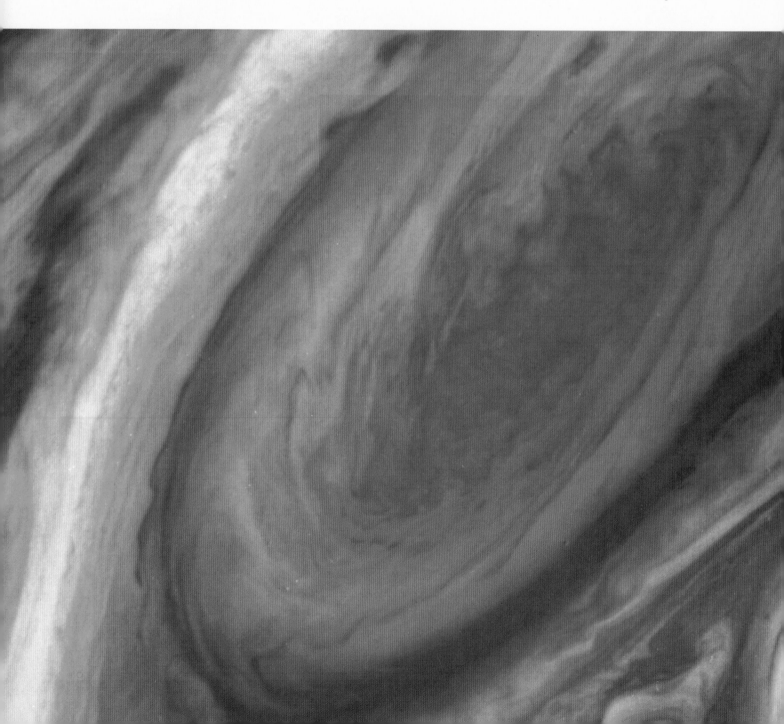

Chapter 5

Looking at the Solar System

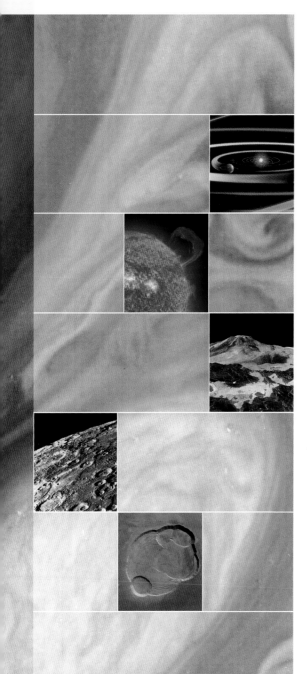

Every morning at dawn, the golden orb of a nearby star rises in the eastern sky and pours light and warmth onto Earth. It is the star we call the Sun.

The Sun is the center of our little corner of the universe. It travels through space along with a motley collection of other bodies big and small. We call the Sun and this family of fellow travelers the Solar System. ("Sol" is the Latin word for Sun.)

It was not always so. Five billion years ago there was nothing in our little corner of space but a tenuous cloud of gas and dust—a nebula. Then, something made the nebula start to shrink, or condense, under gravity. The denser center of this shrinking mass began to heat up and form a spherical shape. Around it, a disk of cooler matter built up. Over time, soaring temperatures inside the sphere triggered off nuclear reactions that set it shining—as the Sun.

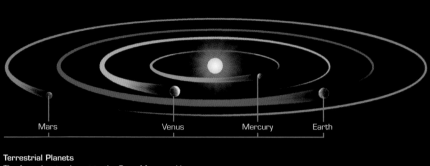

Sun

At the center of the Solar System, the Sun is quite a different body from the planets. It is a star, which gives off light, heat, and other radiation. It is the only body in the Solar System to generate light. All the other bodies shine only because they reflect sunlight. The Sun is just like the other stars we see in the night sky but very much closer. A beam of light from the Sun takes about eight minutes to reach us. A beam of light from even the nearest stars takes more than four years.

Mars   Venus   Mercury   Earth

Terrestrial Planets

The four planets closest to the Sun—Mercury, Venus, Earth, and Mars—are made up mainly of rock, like Earth, which is why we call them the terrestrial (Earthlike) planets. They lie relatively close together compared with the other planets.

Uranus   Neptune   Jupiter Saturn

Gas Giants

The next four planets beyond Mars—Jupiter, Saturn, Uranus, and Neptune—are giant bodies that are made up mainly of gas and liquid gas. That is why we call them gas giants. The main gases present are hydrogen and helium, the most common gases in the entire universe.

Asteroids

Between the terrestrial planets in the inner part of the Solar System and the gas giants in the outer part is a ring of small bodies called the asteroids, or minor planets. They occupy a region known as the asteroid belt. They range in size from less than a mile to several hundred miles wide.

The powerful radiation the newborn Sun gave off started to blow away the light gases from the surrounding disk of matter, leaving dusty particles. Gradually, gravity pulled the particles together to form larger and larger clumps. Eventually, growing ever larger in collisions, the clumps formed into the bodies that became the planets.

The bodies closest to the Sun were made up mainly of rock because the gases that were once present in the warm, inner part of the disk had been blown away. Today, we call these rocky bodies the terrestrial (Earthlike) planets.

In the colder, outer regions of the disk, where solar radiation was weaker, the gases remained and were gathered up by the planetary bodies to form the huge planets we now call gas giants.

Many bits of debris remained after the formation of the planets. And this debris is still present today. There is a ring of small rocky bodies between the terrestrial and gas giants that we call the asteroids. And there appears to be a huge spherical region of icy lumps far beyond the planets that we see as comets when they travel in toward the Sun.

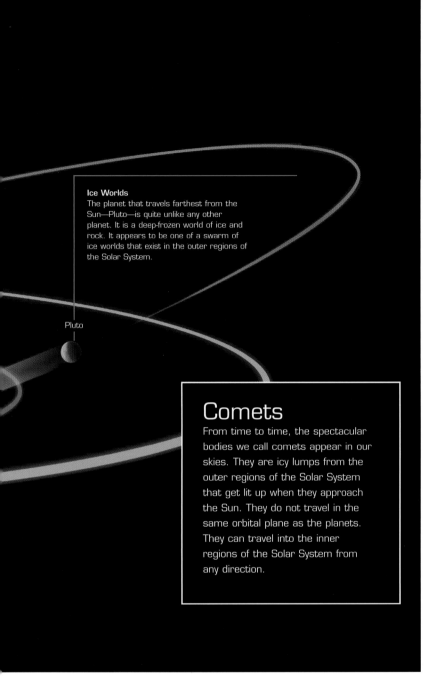

Ice Worlds
The planet that travels farthest from the Sun—Pluto—is quite unlike any other planet. It is a deep-frozen world of ice and rock. It appears to be one of a swarm of ice worlds that exist in the outer regions of the Solar System.

Pluto

Comets

From time to time, the spectacular bodies we call comets appear in our skies. They are icy lumps from the outer regions of the Solar System that get lit up when they approach the Sun. They do not travel in the same orbital plane as the planets. They can travel into the inner regions of the Solar System from any direction.

The Solar System

Astronomy has been flourishing for at least 5,000 years. But the idea of a Sun-centered, or Solar System, did not become accepted until about four hundred years ago. Hitherto, most people believed that Earth was at the center of things: the Sun, the Moon, the planets, and all the heavenly bodies circled around Earth, they said. All this changed after Nicolaus Copernicus realized that the heavenly bodies circle around the Sun (see page 12).

The Sun is a much more worthy contender to be at the center of things. It is more than 100 times wider

than Earth and is 750 times more massive than all the other bodies in the Solar System put together. Because of its great mass, the Sun has an enormous gravitational pull, and it is this invisible force that literally holds the Solar System together. This force extends probably as far as 6 trillion miles (10 trillion km), or 1 light-year. This is nearly a quarter the way to the next nearest star, Proxima Centauri.

Orbits of the Planets

The most important bodies in the Solar System after the Sun are the nine planets, which circle the Sun at different distances. In order, going away from the Sun, they are Mercury, Venus, Earth, Mars, Jupiter, Saturn, Uranus, Neptune, and Pluto. Mercury at times travels as close to the Sun as 28 million miles (45 million km), while Pluto at times travels as far out as 4,600 million miles (7,400 million km).

To say that the planets "circle" the Sun is not quite accurate. Rather, the planets travel in elliptical (oval) paths, or orbits, around the Sun. Mercury and Pluto have the most elliptical orbits.

In their orbits, the planets all travel in the same direction in space. If they could be viewed from a point in space far above Earth's North Pole, the planets would appear to travel in an anticlockwise direction.

The planets also mainly travel in the same plane (flat sheet) in space. This reflects their origin, in which they formed out of a disk that surrounded the newborn Sun. Only Pluto strays appreciably from the general orbital plane.

In a Spin

As well as traveling in orbit around the Sun, the planets have another motion—they spin around on their axis (an imaginary line through their center). Earth spins around once in 24 hours—this defines the time period we call our day. The other planets spin around in different times, called their rotation periods. Venus spins around slowest, taking over 243 days; Jupiter spins around fastest, in under 10 hours.

In the case of Jupiter, the axis around which it spins is nearly upright, or perpendicular, in relation to the direction it is traveling in its orbit. But most other planets spin around an axis that is tilted at an angle

to the direction in which it is traveling. Earth's axis, for example, is tilted 23½ degrees out of the perpendicular. It is this tilt that causes the seasons. When a particular place on Earth is tilted more toward the Sun, it is warmer, but when it tilts more away, it becomes less warm and the season changes.

Other Solar Systems

Astronomers have long suspected that many other stars besides the Sun have planets orbiting around them. But it was not until 1991 that the first such planets were discovered, circling around a tiny neutron star (see page 35). Four years later, a planet was discovered circling round an ordinary star, 51 Pegasi.

Since that time, more than 100 planets have been found circling around ordinary stars. They are known as extrasolar planets. On average, they seem to be roughly the size of Jupiter. They are too small to be spotted in telescopes, and are detected indirectly. They make the stars they are circling wobble slightly, and this wobble can be detected.

In the Orion Nebula alone, the Hubble Space Telescope has spotted hundreds of dusty disks around newborn stars that may one day turn into planets. They are known as protoplanetary disks. It seems as if planet formation is common in the universe.

Planets in the Sky

At times, we can see some of the planets shining in the night sky like bright stars. But, unlike the true stars, the planets are always changing their positions among the stars of the constellations. Ancient astronomers were familiar with these "wandering stars" and called them planets, after the Greek word for "wanderer."

Venus is the most familiar of these wandering stars. Sometimes we can see it shining brilliantly in the western sky just after sunset, when we call it the evening star. At other times we can see it as a morning star shining in the east just before sunrise. We can see Mercury too as an evening or morning star, but it is more difficult to find because it always stays much closer to the horizon than Venus.

Mars varies in brightness depending on its location in its orbit. At its farthest point from Earth, it is not easy to find. But when it is closest to us (at opposition), it is brighter than all the stars and unmistakable because of its fiery red color. This is why we also call Mars the Red Planet.

Jupiter can appear equally bright but can be distinguished from Mars because of its brilliant white color. Saturn can also outshine most stars at times, but can be difficult to find when it is farthest from us.

Saturn is the farthest planet easily visible to the naked eye, and the most distant known to early astronomers. They could not see the three other planets discovered in more recent times—Uranus, Neptune, and Pluto.

The Terrestrial Planets

The terrestrial or Earthlike planets—Mercury, Venus, Earth, and Mars—are quite different in size and composition from the gas giants farther out. They broadly have a similar structure, made up of layers of rock with a metal center, or core.

The uppermost layer, made up of hard rock, is called the crust. Underneath it is a more dense rock layer known as the mantle. This overlies the core, which is made up primarily of iron.

Among the four terrestrial planets, Venus and Earth are surrounded by a relatively thick layer of gases, which form an atmosphere. Mercury has virtually no atmosphere at all, while that of Mars is very thin indeed. The actual surface conditions on each terrestrial planet are quite different, and are partly dependent on the presence—or absence—of an atmosphere.

Landscapes

Mercury has a rugged, heavily cratered surface. The landscape of Venus, however, has few craters and is dominated by volcanoes and vast lava plains. Earth has the most varied landscape of all the terrestrial planets, shaped by bodily movements in the crust, the action of the weather, and flowing water in vast oceans and rivers.

Mars is essentially a great desert region, heavily cratered in some parts and dominated by huge volcanoes in others.

Moons

The terrestrial planets have few moons between them. Mercury and Venus don't have any moons. Earth has just one—the Moon, which, as moons go, is relatively large compared with the size of its parent planet.

Mars has two moons, called Phobos and Deimos, but they are tiny. They are probably asteroids that Mars captured long ago when they strayed too close to the planet from the asteroid belt.

The Asteroids

The next planet after Mars in the Solar System, going away from the Sun, is Jupiter. This planet orbits more than 340 million miles (550 million km) farther away than Mars. But the space between Mars and Jupiter is not empty. It is home to the asteroids, which orbit the Sun in a broad band, or belt, about 200 million miles (300 million km) wide.

The asteroids seem to be lumps left over after the formation of the planets. They couldn't combine to form a planet themselves because of the powerful gravitational disturbances caused by nearby Jupiter. There are billions of asteroids, but only a handful are relatively large. Biggest and first to be discovered (in 1801) is Ceres, which is about 580 miles (930 km) across. Pallas (diameter about 400 miles/600 km) and Vesta (350 miles/550 km) are the next largest. But most asteroids measure much less than 60 miles (100 km) across. Vesta, incidentally, is the only asteroid that can be glimpsed, under ideal conditions, with the naked eye.

Gas Giants

The asteroids serve as a kind of boundary between the four terrestrial planets and the next four planets out— Jupiter, Saturn, Uranus, and Neptune. These four planets are appropriately called the gas giants. They are epitomized by Jupiter, which is truly gigantic. It is 11 times as big across as Earth, and has more than 300 times Earth's mass. In the Solar System, only the Sun is bigger. The gas giants have similar basic structures. Unlike the terrestrial planets, they have no solid surface. They have a very deep atmosphere, mainly of hydrogen and helium. Beneath the atmosphere there is a vast planet-wide ocean.

Oceans

In the case of Jupiter and Saturn, the ocean consists of liquid hydrogen and helium. The pressure of the deep atmosphere has compressed the gases into a liquid state. Many thousands of miles beneath the ocean surface, pressures build up so much that they compress hydrogen into a kind of liquid metal (perhaps like the liquid metal mercury we find on Earth). Only at the center of the planet is there probably a solid, rocky core, several Earth's widths across.

The smaller gas giants Uranus and Neptune—each about four times as big across as Earth—have a different kind of ocean, made up of water and liquid gases such as ammonia and methane.

Many Moons

The gas giants also have two other things in common—they each have an abundance of moons and are surrounded by a system of rings. Whereas the terrestrial planets have only three moons among them, the four gas giants have more than 130, and more are being discovered all the time. Jupiter's moon, Ganymede, and Saturn's moon, Titan, are each larger than the planet Mercury.

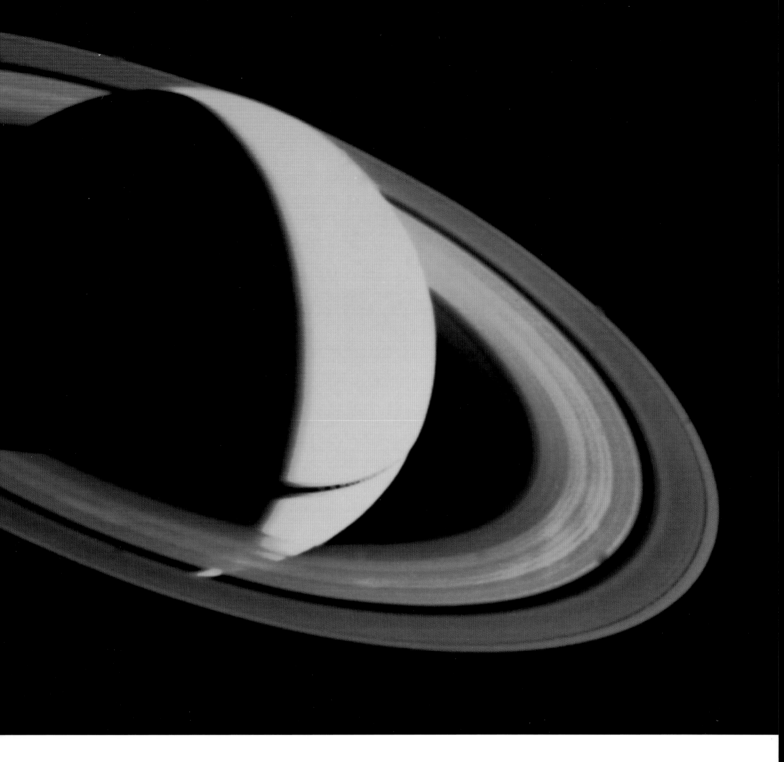

Most of the moons of the giant planets are peppered with craters like our own Moon, but they have an icy surface. Jupiter's Io is unique because it has active volcanoes that spew out sulfur onto the surface. Saturn's Titan is unique because it has a dense atmosphere.

Rings

Once it was thought that only Saturn was surrounded by a ring system, but close-up investigation by the *Voyager* probes has revealed that the other three gas giants also have rings.

Saturn's rings are by far the most prominent. Shining brilliantly, they make Saturn a most glorious sight in telescopes. The rings of the other giants are much darker and fainter, and invisible from Earth.

Sun

The Sun is a great globe of intensely hot, seething gases that pours out light, heat, and other radiation into space. Its bright surface, called the photosphere ("light-sphere"), has a temperature of around 10,000°F (5,500°C).

But in the Sun's core (center), temperatures rise to 27,000,000°F (15,000,000°C) or more. At such temperatures, nuclear reactions take place that fuse hydrogen nuclei into helium and produce the fantastic amounts of energy that keep the Sun shining.

Seen close up, the Sun's surface has a speckled appearance. This so-called granulation is caused by pockets of hot gas rising, cooling, and then descending. From time to time, darker patches appear on the photosphere, called "sunspots," which are about 2,700°F (1,500°C) cooler than their surroundings. They are caused by the presence of intense magnetic fields.

Above the photosphere, the Sun has a thinner inner atmosphere called the chromosphere (color-sphere), about 3,000 miles (5,000 km) deep. This gives way to an even thinner outer atmosphere known as the corona (crown), which gradually merges into space.

This image, showing the upper part of the chromosphere, was returned by SOHO (the Solar and Heliospheric Observatory) at ultraviolet wavelengths. It is a region of intense activity. The bright areas signal where violent explosions, which we call solar flares, are taking place.

But the most spectacular feature is the great eruptive prominence at top right. Prominences are fountains of cooler (100,000°F/60,000°C) gas that leap into the intensely hot corona, which typically has a temperature of over 1,800,000°F (1,000,000°C). ➤

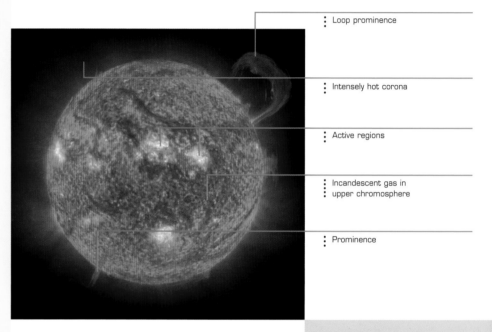

- Loop prominence
- Intensely hot corona
- Active regions
- Incandescent gas in upper chromosphere
- Prominence

Looping the Loop

A huge solar prominence punches its way into space, extending hundreds of thousands of miles above the Sun's surface. It follows the contours of the Sun's powerful magnetic field.

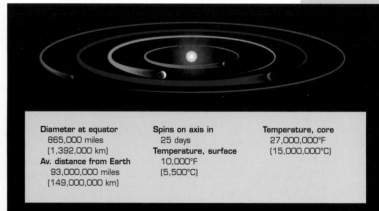

Diameter at equator	Spins on axis in	Temperature, core
865,000 miles (1,392,000 km)	25 days	27,000,000°F (15,000,000°C)
Av. distance from Earth 93,000,000 miles (149,000,000 km)	**Temperature, surface** 10,000°F (5,500°C)	

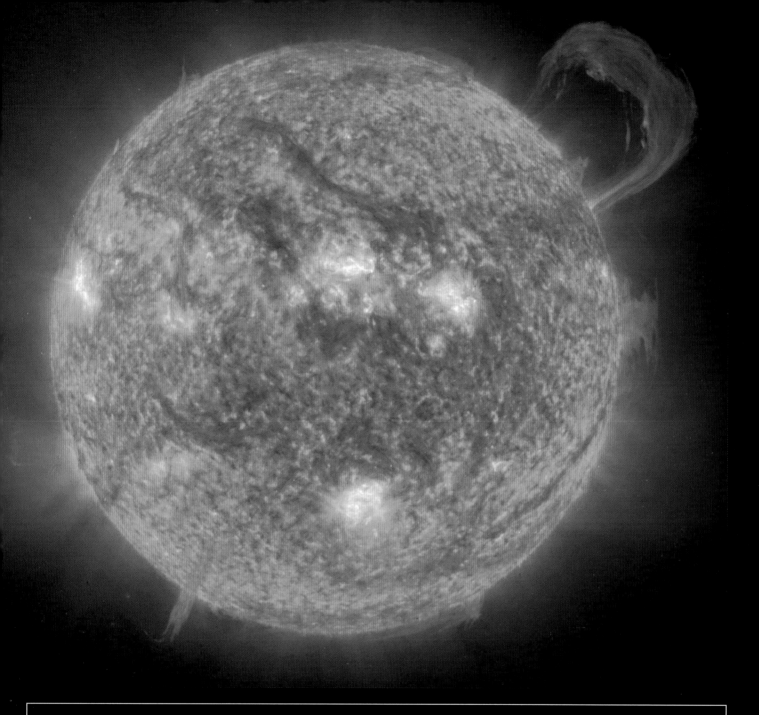

Sunspots

Early in 2001, a huge sunspot group developed on the Sun's surface, pictured here by SOHO. It spanned a region 13 times the area of Earth's surface. It was the source of powerful coronal mass ejections and the biggest solar flare seen for a quarter century. Sunspots come and go according to a regular solar cycle of about 11 years.

Approx size of Earth ⟶ •

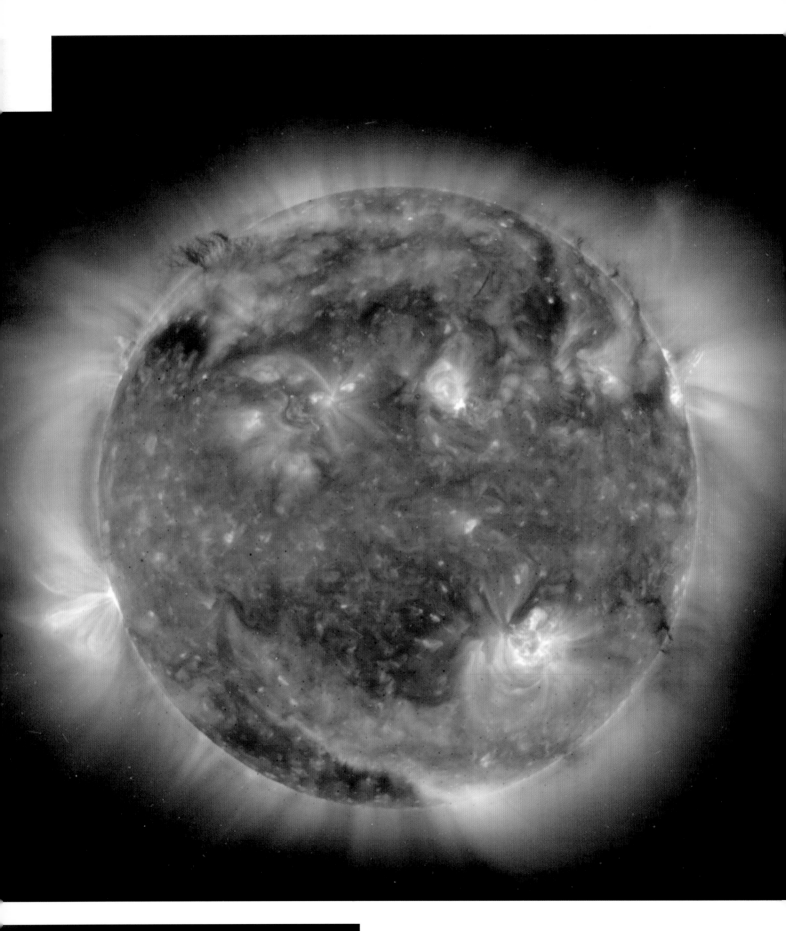

n Full Color

One of the most amazing pictures of the Sun, derived from images returned by SOHO. Produced by combining images taken at three wavelengths, it portrays the Sun as a vibrant, seething entity, pouring out waves of radiation and particles that light up the abysmal darkness of the surrounding space.

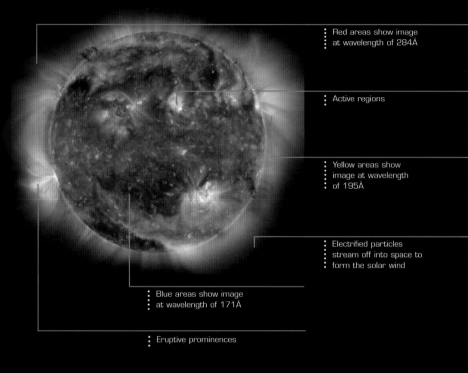

Red areas show image at wavelength of 284Å

Active regions

Yellow areas show image at wavelength of 195Å

Electrified particles stream off into space to form the solar wind

Blue areas show image at wavelength of 171Å

Eruptive prominences

Coronal Loops

Loops of superhot electrified gas leap from the solar surface into the corona, the Sun's outer atmosphere. They carry the energy that makes temperatures in the corona soar beyond 1,800,000°F (1,000,000°C). The spacecraft, TRACE (Transition Region and Coronal Explorer), returned this stunning image showing clusters of coronal loops 300,000 miles (500,000 km) high.

➤ Astronomers from Galileo on have studied the Sun enthusiastically. Astronomers on the ground study the Sun with solar telescopes, like the 11-story-high McMath telescope at Kitt Peak Observatory in Arizona, the world's biggest.

But it is from observations from space that we have gained most knowledge about our local star. The U.S. space station *Skylab* pioneered this study in 1973, observing the Sun continuously for periods of months at a time. The satellite *Solar Max* (launched 1980) and probe *Ulysses* (1990) followed.

However, it has been SOHO (the Solar and Heliospheric Observatory), launched in 1995 and still going strong, that has revealed the Sun as never before. It observes the Sun from one of the so-called Lagrangian points in Earth's orbit, some 930,000 miles (1,500,000 km) from Earth.

SOHO observes the Sun continuously with a set of 12 main instruments, provided by solar researchers in the United States and throughout Europe. They monitor the day-to-day activity on the Sun, recording solar flares and sunspots, coronal mass ejections—the outflow of energetic particles from the corona—and the varying strength of the solar wind.

The Extreme Ultraviolet Imaging Telescope (EIT) has returned some of the most spectacular images of the Sun. The large picture here is a stunning example. It is a composite picture made by combining images taken at three different ultraviolet wavelengths—171Å, 195Å, and 284Å. (Å is the symbol for the Angstrom unit, which is equal to one ten-billionth of a meter.)

Eclipses

It is one of the strangest coincidences in nature. The Sun is about 400 times bigger across than the Moon, but it lies about 400 times farther away. As a consequence, both the Sun and the Moon appear much the same size in our skies.

From time to time, as the Moon orbits Earth and Earth orbits the Sun, the Moon passes in front of the Sun, as seen from Earth. Because the Sun and Moon are the same size, the Moon blots out the Sun's light. We call this an eclipse of the Sun, or solar eclipse. At other times, Earth comes between the Sun and the Moon and casts a shadow on the Moon. We call this an eclipse of the Moon, or lunar eclipse. Both solar and lunar eclipses can be partial, when only part of the Sun or Moon is covered, or total, when the entire Sun or Moon is covered.

Our main picture shows what happens during a total lunar eclipse. The Moon enters Earth's shadow but does not completely disappear from view. This is because a certain amount of sunlight reaches the Moon after being refracted (bent) by Earth's atmosphere. A total lunar eclipse can last for up to about two-and-a-half hours.

Total solar eclipses, on the other hand, can last for a maximum of only about seven-and-a-half minutes. Usually, they are much shorter. This is because the Moon casts only a tiny shadow on Earth—never more than about 170 miles (270 km) across. And this shadow races over the surface at a speed of over 1,000 mph (1,600 km/h) as the Moon travels in its orbit.

Brief they may be, but total solar eclipses are among the most spectacular events in nature. During the eclipse, the air chills as day turns suddenly into night; birds start to roost and flowers close. Astronomers "chase" eclipses the world over. They can be rewarded by viewing the solar chromosphere, the corona, and prominences—or they may be clouded out!

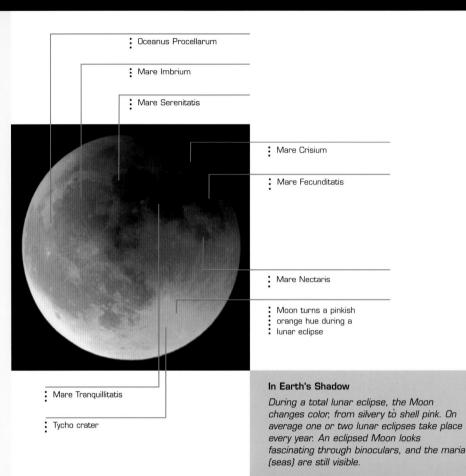

- Oceanus Procellarum
- Mare Imbrium
- Mare Serenitatis
- Mare Crisium
- Mare Fecunditatis
- Mare Nectaris
- Moon turns a pinkish orange hue during a lunar eclipse
- Mare Tranquillitatis
- Tycho crater

In Earth's Shadow

During a total lunar eclipse, the Moon changes color, from silvery to shell pink. On average one or two lunar eclipses take place every year. An eclipsed Moon looks fascinating through binoculars, and the maria (seas) are still visible.

Total Solar Eclipse

The time of totality during the total eclipse of the Sun that occurred on July 11, 1991, viewed from Hawaii. It was one of the longest eclipses of the twentieth century. Note the colored chromosphere visible around the limb of the Moon, and the prominences visible at 12 o'clock and 6 o'clock. It is only during total eclipses that we are able to see the chromosphere and prominences from Earth.

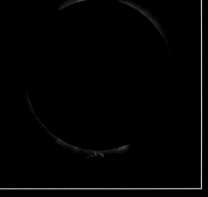

Moon

The Moon is Earth's nearest neighbor in space, its constant companion. We can see the Moon in the sky most nights of the year, changing in appearance from a slim crescent to full circle, and back again. We call these changing shapes its phases. The phases come about because we see more or less of the Moon's face illuminated by the Sun as the Moon travels in orbit around Earth once a month. The Moon, of course, has no light of its own but shines by reflected sunlight.

When we see the whole face of the Moon lit up, we call it the full moon. The full moon we see in the picture, however, is not the full moon we see from Earth. It is a full-circle view of the Moon seen from a different perspective by the *Apollo 11* astronauts when they traveled to the Moon in July 1969.

The left half of the image shows what we see on the right half of our full moon. (Compare with our eclipsed full moon on page 132.) The right half of the image shows part of the hidden side of the Moon, which we can never see from Earth.

The large dark areas in the left part of the image are the regions we still call maria, or seas, even though we know they are dry, dusty plains. Maria dominate the face of the Moon we see from Earth. By contrast, no large mare regions are found on the Moon's hidden side, as this image confirms. ➤

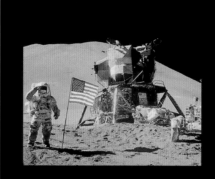

Man on the Moon

Astronaut James Irwin salutes the Stars and Stripes at the start of the *Apollo 15* Moon-landing mission to the foothills of the Apennine mountain range, which forms the boundary between Mare Imbrium and Mare Serenitatis. Behind him is the lunar landing module, and at right is the lunar rover—Moon buggy—being used for the first time. Towering in the background is Mount Hadley. The *Apollo 15* mission took place between July 26 and August 7, 1971.

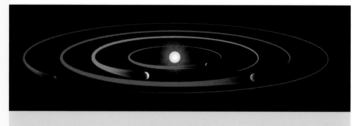

Diameter at equator	Av. distance from Earth	Spins on axis in
2,160 miles	239,000 miles	27.3 days
(3,476 km)	(384,000 km)	**Goes through phases in**
	Circles Earth in	29.5 days
	27.3 days	

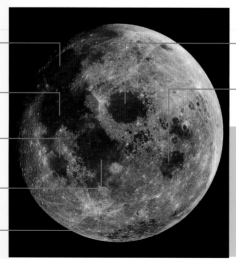

Mare Serenitatis
(Sea of Serenity)

Mare Tranquillitatis
(Sea of Tranquility)

Mare Nectaris
(Sea of Nectar)

Mare Fecunditatis
(Sea of Fertility)

Cratered highlands

Mare Crisium
(Sea of Crises)

Far side of Moon

Seas and Highlands

Even with our eyes alone we can see dark and light areas on the Moon. The dark areas are the great flat plains we call maria (seas) and the bright areas are highlands. The maria are younger than the more heavily cratered highlands.

: Crater Herschel

: Crater Flammarion

: Central peaks

: Edge of Oceanus
: Procellarum
: (Ocean of Storms)

: Smooth crater floor

: Crater Ptolemaeus

: Dimple craters

: Crater Lalande

Lunar Landscape

A beautiful lunar landscape, close to the center of the Moon. Ptolemaeus is one of a chain of large craters that extends almost due south from the center. They show up best at the "half-moon" phases of first quarter and last quarter.

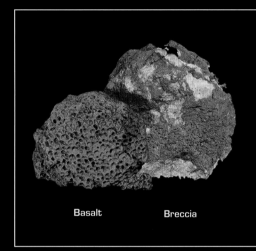

Basalt   **Breccia**

Moon Rocks

Two typical kinds of rock found on the Moon and carried back to Earth by the *Apollo* astronauts. Altogether, the astronauts returned with a total of about 850 pounds (385 kg) of rock and soil samples. Lunar basalt is similar to the rock found around volcanoes on Earth. This sample is riddled with holes where gases escaped from the molten rock. Lunar breccia is a rock made up of rock chips cemented together.

➤ Although the Moon has no blue sky, luxuriant greenery, or colorful flowers and no rippling streams or azure oceans, it has an ethereal beauty all its own, captured here in an image taken by *Apollo* astronauts. It pictures a landscape just south of a small mare region at the Moon's center called Sinus Medii.

As elsewhere on the Moon, the surface is peppered with craters large and small. The tiniest, just a few yards across, are called dimple craters. The largest can be hundreds of miles across. The largest one in the picture, Ptolemaeus, has a diameter of 90 miles (145 km). It has relatively low walls and a flat, smooth floor. Such craters are sometimes called walled plains.

The small crater next to it, Herschel, is much smaller, only 28 miles (45 km) across. It has the classic profile of many lunar craters—nearly circular with terraced walls and a central mountain range.

Near the bottom of the picture, the small crater Lalande (25 miles/40 km across) illustrates the basic characteristics of a crater—the floor descends well below, and the walls rise well above, the surrounding lunar surface.

Most of the craters found on the Moon have resulted from the impact of rocks from outer space. Relatively few have been formed by volcanoes. In the early days of the Moon's

history, space was full of lumps of rock, large and small. They bombarded the Moon's surface mercilessly to create the heavily cratered landscape of the lunar highlands.

When particularly large space rocks rained down, they gouged out great basins in the Moon's crust, which subsequently filled with lava. This is the origin of the lunar maria, or seas. They are great lava plains, surrounded by the remnants of the walls of the original craters.

Mercury

Mercury is the planet that is closest to the Sun, and observational astronomers find Mercury a difficult subject because it always remains close to the Sun in the sky. Occasionally they can glimpse it as an evening or morning star close to the horizon. But only vague markings can be seen on its surface, even with a powerful telescope.

It was not until March 1974 that astronomers were treated to their first close-up view of Mercury's surface. This was provided by the probe *Mariner 10*, which had journeyed to the planet via Venus. *Mariner's* cameras returned images that revealed that Mercury is covered in craters and looks similar to the heavily cratered regions of the Moon.

In general, the craters on Mercury are shallower than those on the Moon because of the planet's stronger gravity. Also, Mercury doesn't have the extensive mare (sea) regions that the Moon has, but it has smoother regions between the craters known as intercrater plains.

Mariner 10 flew past Mercury three times. On the second encounter, in September 1974, it returned the image we see here, which shows part of the south polar region. The largest crater at left measures around 90 miles (150 km) across. It has been ruined (altered) by the impact of a more recent crater. This one shows the same characteristics as many large lunar craters—terraced walls and central mountain peaks.

Astronomers will have to wait until March 2011 before they can see the whole of Mercury's surface. That is when NASA's *Messenger* spacecraft (launched August 2004) is scheduled to begin orbiting the planet. *Messenger* by then will have flown past Earth, Venus (twice), and Mercury itself (three times) on a circuitous, minimal-energy trajectory to its target. (*Messenger* stands for Mercury surface, space environment, geochemistry, and ranging.)

Crater world

A typical Mercurian landscape, peppered with craters large and small. The older craters have been "ruined" by the impact of younger ones. In between the larger craters there are smoother regions known as intercrater plains.

Diameter at equator	Av. distance from Sun	Spins on axis in
3,032 miles	36,000,000 miles	58.6 days
(4,880 km)	(58,000,000 km)	**Number of moons**
	Circles Sun in	0
	88 days	

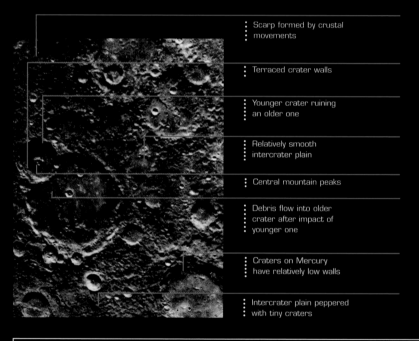

Scarp formed by crustal movements

Terraced crater walls

Younger crater ruining an older one

Relatively smooth intercrater plain

Central mountain peaks

Debris flow into older crater after impact of younger one

Craters on Mercury have relatively low walls

Intercrater plain peppered with tiny craters

Farewell Images

As *Mariner 10* retreated from Mercury on March 29, 1974, it snapped a series of 18 images, which have been combined here into a photomosaic. This shows a heavily cratered landscape, but notice that there are smoother (and darker) regions reminiscent of the mare (sea) areas on the Moon. In the upper right of the image, there are extensive ray patterns around some of the craters, showing where matter was ejected when the craters were formed. Most of the landscape remains much as it was following intense bombardment by space rocks four billion years ago.

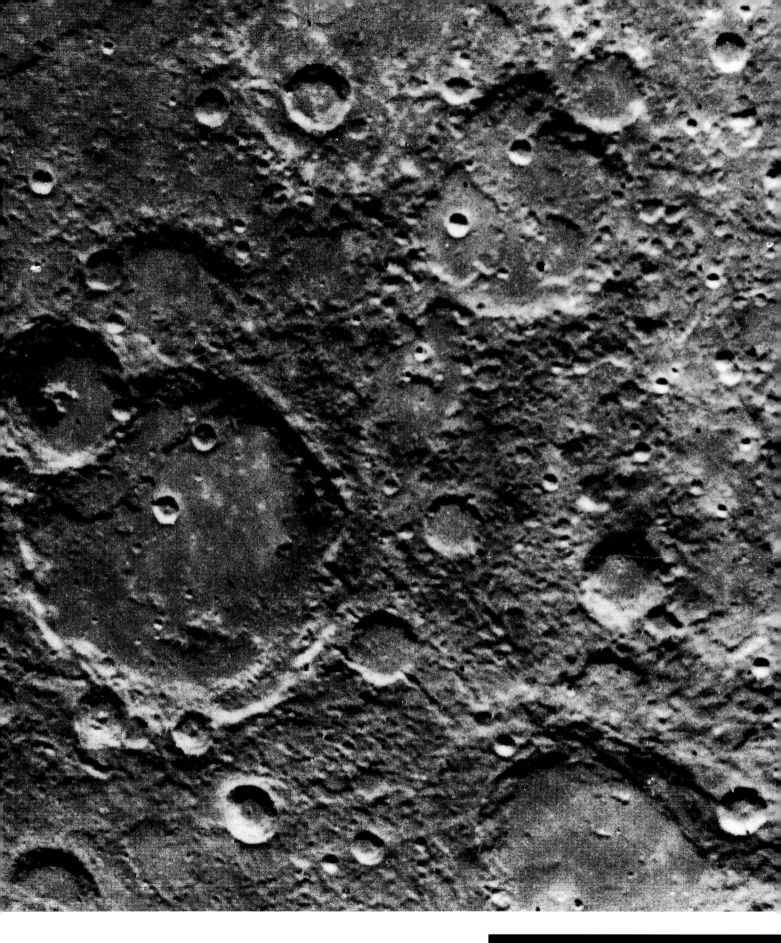

Venus

Venus is the planet that comes closest to Earth—within 26 million miles (42 million km). We can see it on many evenings of the year shining brightly in the western sky just after sunset—as the evening star.

Like Earth, Venus has an atmosphere and is a near twin of Earth in size. But the two planets could hardly be more different. Venus's atmosphere is very much denser than Earth's and is made up mainly of carbon dioxide—notorious as the major "greenhouse" gas on Earth that leads to global warming. On Venus, there has been a runaway greenhouse effect that has pushed temperatures on the planet as high as 600°F (480°C).

Venus's atmosphere is full of thick clouds (of sulfuric acid droplets) that prevent us from seeing the surface underneath. But we now know what the surface is like, thanks to orbiting probes that have scanned the surface with radar—radar uses radio waves, which can penetrate cloud. The U.S. probe *Magellan* spent nearly four years scanning Venus, between 1990 and 1994. It mapped virtually the whole planet.

The picture here is a composite, based on multiple images returned by *Magellan*. It shows a hemisphere centered on one of the two main highland regions, or "continents" on Venus, named Aphrodite Terra. It is located near the planet's equator. The other continent, named Ishtar Terra, is located out of the picture close to the North Pole. Ishtar Terra is about the size of Australia; Aphrodite Terra, the size of Africa.

The brightest parts of the image show highland regions, which include areas such as Thetis Regio and Alta Regio. The highland areas are crisscrossed in many places by deep rifts, or valleys, such as Artemis Chasma and Diana Chasma. Such rifts can run for hundreds of miles and can be many miles deep.

But most of Venus consists of lowland plains, named planitia. They have been formed by the outflow of lava from the hundreds of volcanoes that dot Venus's surface.

Global Venus

A global view of Venus, based on images returned by the Magellan *probe. It is centered on one of the planet's two "continents," Aphrodite Terra. Planitia are plains regions. Regio is a highland, chasma is a valley.*

Diameter at equator	Av. distance from Sun	Spins on axis in
7,521 miles	67,000,000 miles	243 days
(12,104 km)	(108,000,000 km)	**Number of moons**
	Circles Sun in	0
	224.7 days	

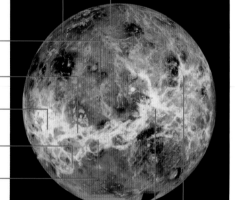

- Niobe Planitia
- Atalanta Planitia
- Rusalka Planitia
- Diana Chasma
- Thetis Regio
- Artemis Chasma
- Aphrodite Terra
- Alta Regio
- Ulfrun Regio

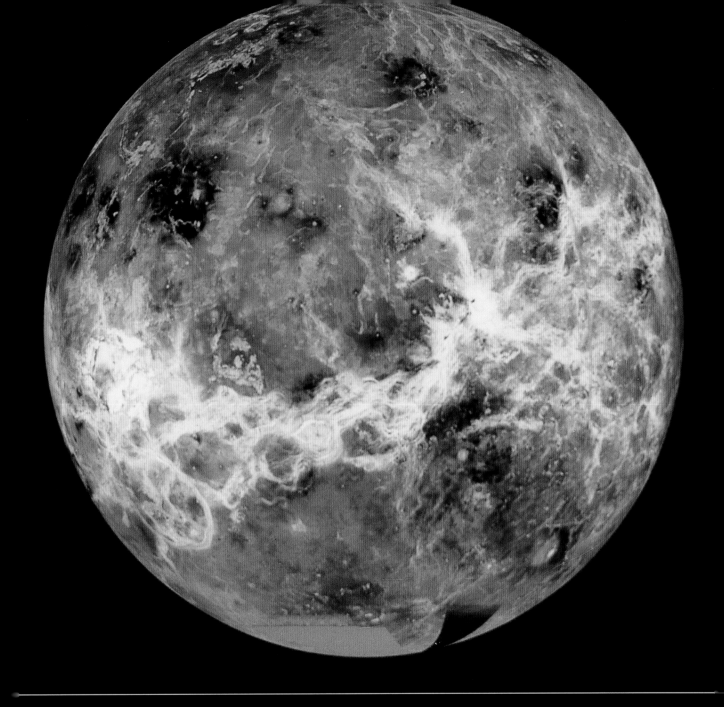

Maat Mons

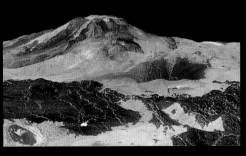

Named after an Egyptian goddess, Maat Mons is one of the most imposing volcanoes on Venus. It rises to a height of some 6 miles (9 km) from a base 125 miles (200 km) across. The picture shows vividly the lava flows that have cascaded down the slopes and spread themselves over the surrounding landscape. Some volcanoes on Venus measure more than 250 miles (400 km) across. Only on Mars have bigger volcanoes been found in our Solar System.

Earth

Our home planet, Earth, is the archetypal terrestrial planet, made up mainly of rock. It is larger than the others—Mercury, Venus, and Mars—and differs from them in one important respect. Its surface is ever-changing because of erosion by flowing water in the rivers and oceans, and by the weather. These agencies continually attack the surface rocks, in time reducing the loftiest mountains into gentle hillocks.

But more fundamental changes take place beneath Earth's crust (top layer), in the upper part of the rocky layer underneath, called the mantle. The upper-mantle rocks are hot and semi-molten, and on the move. They carry with them sections, or plates, of the crust. The plates move in different directions. Some collide and throw up mountain ranges. Others move apart, which is happening, for example, in the Atlantic Ocean and causing the continents of Europe and North America— which each sit on different plates—to drift apart.

This picture of Earth was taken by the weather satellite *Meteosat* from an altitude of around 22,400 miles (36,000 km). It shows particularly cloud-free land masses of Africa and Europe. It is more cloudy in the north over the Arctic and in the south above the Antarctic ice cap. Prominent is the second-largest ocean on Earth, the Atlantic, which covers an area nearly ten times that of the United States. Overall, oceans cover more than 70 percent of Earth's surface.

At left is the continent of South America, the eastern side of which would have fitted into the western side of Africa 200 million years ago. This was the time when all the present-day continents were merged into one supercontinent, Pangaea. Then continental drift began splitting this supercontinent apart, resulting in the distribution of the continents we find today. Two hundred million years hence, Earth's surface will look completely different again.

Active Earth

Molten lava erupts from a volcano in Hawaii and flows over the surrounding landscape. There are hundreds of volcanoes active on Earth, showing that geologically, our planet is very much alive. Most volcanoes are found at the edges of the plates, or moving sections of crust, that form Earth's relatively mobile surface. No other planet is as geologically active. They seem to be long-dead worlds.

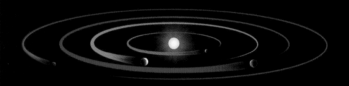

Diameter at equator	Av. distance from Sun	Number of moons
7,926 miles (12,756 km)	93,000,000 miles (149,600,000 km) **Circles Sun in** 365.25 days	1 (the Moon)

Home Planet

Photographed from space, Earth is a colorf[...] world, dominated by the deep blue of the oceans[...] White flecks of cloud ar[...] scattered throughout th[...] atmosphere. Its near-transparency allows us to view the land areas, or continents.

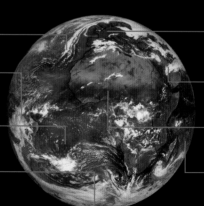

Clouds in atmosphere

Continent of South America

Atlantic Ocean

Continent of Antarctica

Continent of Europe

Arabian Peninsula

Continent of Africa

Indian Ocean

Mars

Astronomers look forward to the times when Mars comes closest to Earth. This happens when the planet is at opposition, or on the opposite side of Earth from the Sun. Oppositions occur every 26 months, but, because of Mars's eccentric orbit, some oppositions are more favorable than others.

The opposition on August 27, 2003, was the most favorable in 59,619 years. Mars came within 34.65 million miles (55.76 million km) of Earth, virtually the closest it can ever get. It shone brilliantly in the night sky that summer, approaching a visual magnitude of -3. Its fiery red hue made it absolutely unmistakable. Mars is indeed the Red Planet.

The Hubble Space Telescope was on hand to record this momentous event, and the two images opposite were the result. They were taken 11 hours apart, during which time Mars had nearly completed a half-turn. So the two images capture features of nearly the whole surface of the planet.

The two views show well the difference between the northern and southern hemispheres. The northern hemisphere is a region of vast deserts that astronomers believe may once have been great oceans. The southern hemisphere is more rugged and more heavily cratered.

In the image at the right, the four great Martian volcanoes are just visible—lofty Olympus Mons and the three nearby volcanoes on the Tharsis Ridge—Ascreus Mons, Pavonis Mons, and Arsia Mons. To their right, running close to Mars's equator, lies the other outstanding geological feature on Mars, the 3,000-mile- (5,000-km-) long Valles Marineris—Mariner Valley. This has been called Mars's "Grand Canyon," but it dwarfs Arizona's Grand Canyon many times over. ➤

Close Encounters

Two views of Mars returned by the Hubble Space Telescope in August 2003, when the planet made its very close approach to Earth. They show the dark markings that have intrigued astronomers ever since they turned their telescopes on Mars.

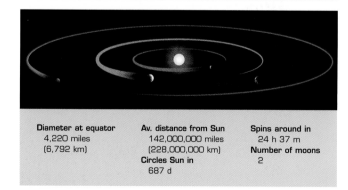

Diameter at equator	Av. distance from Sun	Spins around in
4,220 miles	142,000,000 miles	24 h 37 m
(6,792 km)	(228,000,000 km)	**Number of moons**
	Circles Sun in	2
	687 d	

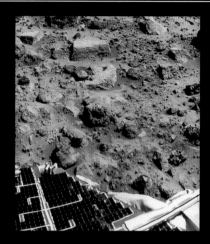

Rugged Rocks

Rocks embedded in rust-colored soil lie scattered around the landing site of the *Pathfinder* probe, which set down on Mars in 1997. *Pathfinder* landed in what was thought to be an ancient river bed at the mouth of Ares Valles (Mars Valley). And some of the rocks showed signs that they had been eroded by flowing water.

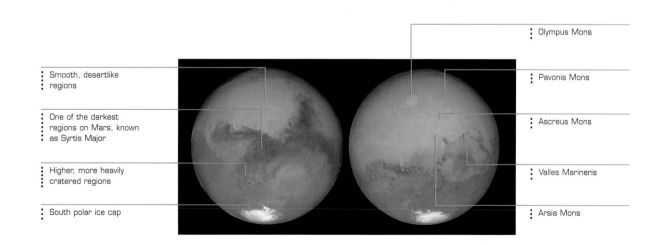

Olympus Mons

Smooth, desertlike regions

One of the darkest regions on Mars, known as Syrtis Major

Pavonis Mons

Higher, more heavily cratered regions

Ascreus Mons

Valles Marineris

South polar ice cap

Arsia Mons

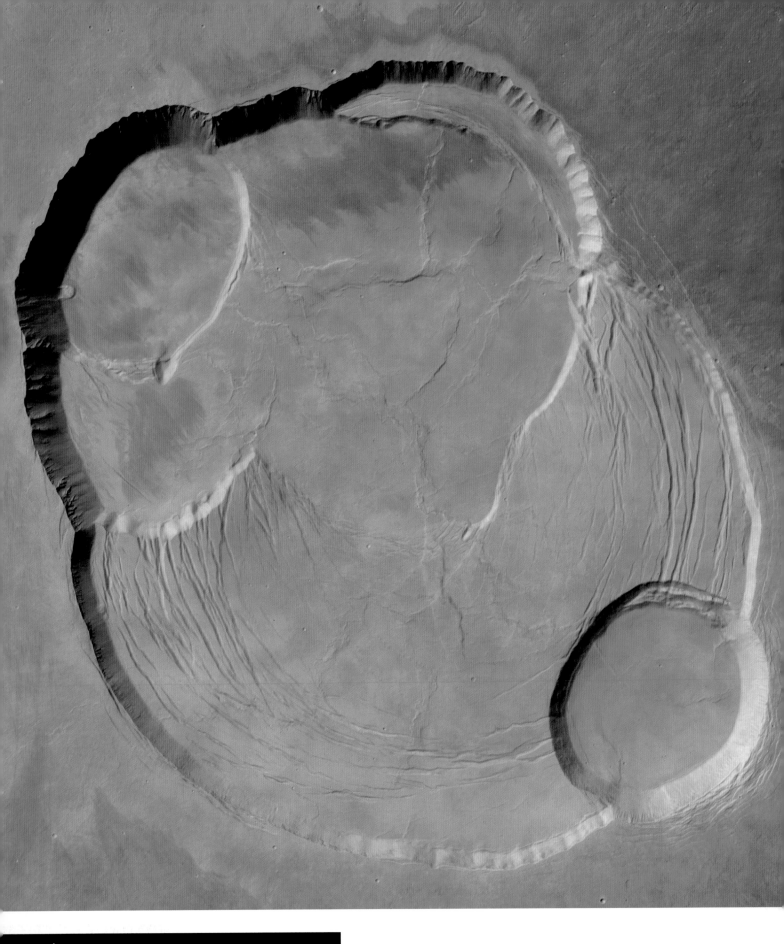

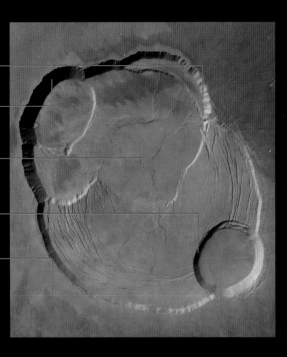

Summit caldera

Average depth of
caldera, about 2 miles
(3 km)

Smooth caldera floor
formed of congealed
lava

Collapsed craters from
multiple eruptions

Gentle slopes of old lava
flows

Complex Caldera

*The caldera, or summit crater, of Mars's highest volcano
Olympus Mons. It is a multiple crater system, indicating
that the volcano has erupted many times. Almost certainly
it is now extinct, like the other Martian volcanoes.*

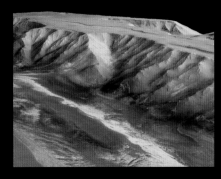

Condor Chasma

This deep canyon, known as Condor
Chasma, is one of the largest canyons
in Mars's "Grand Canyon"—Valles
Marineris. It has been imaged in 3D
by the High Resolution Stereo Camera
on *Mars Express*. Valles Marineris
(Mariner Valley) was named for the
Mariner 9 spacecraft that first
spotted it. The valley is a geological
fault system that runs for a distance
of some 3,000 miles (5,000 km). In
places it is 250 miles (400 km) wide
and 4 miles (6 km) deep.

➤ Much of the northern hemisphere of Mars
is occupied by relatively low-lying, smooth
plains. But at one location near the equator,
Mars's crust suddenly rises to create what is
called the Tharsis Bulge. Topping the Bulge is
a line of three huge volcanoes, rising up to 12
miles (20 km) high. They are Ascreus Mons,
Pavonis Mons, and Arsia Mons; and to their
west lies the mighty Olympus Mons (see page
144).

Olympus Mons—Mount Olympus—is not only
the biggest volcano on Mars, it is also the
biggest we know in the whole Solar System. It
soars to a height of 14 miles (22 km), which
makes it nearly three times the size of Mount
Everest on Earth. At the base, it measures a
staggering 400 miles (600 km) across. It
dwarfs similar volcanoes on Earth, such as
Mauna Loa in Hawaii. The main reason
volcanoes on Mars were able to grow so
much bigger than Earth's is that the planet's
gravity is much lower (38 percent of Earth's).

The picture shows an overhead view of the
summit of Olympus Mons, taken by the High
Resolution Stereo Camera of the orbiting
Mars Express. The probe went into orbit
around Mars early in January 2004 and
began returning some of the best images yet
of Mars's surface.

The summit caldera, or collapsed crater, is
complex in structure, showing evidence of
multiple eruptions. The surrounding surface is
smooth and uncratered, suggesting that the
volcano was active as recently as 150 million
years ago. The gentle slopes surrounding the
summit tell us that Olympus Mons is a type of
volcano known as a shield volcano—like
Mauna Loa. Shield volcanoes spew out very
runny lava that flows a long way.

Jupiter

Eleven times bigger across than Earth, and more massive than all the other planets put together, Jupiter is the largest of the gas giants and by far the biggest planet in the Solar System.

Jupiter presents a colorful view in a telescope. The atmosphere is mainly reddish orange in color and is split into dark and light bands, called belts and zones. The bands are regions of clouds that have been drawn out parallel with the equator because the planet rotates so quickly—once in less than ten hours.

The picture shows a close-up of the most fascinating feature in Jupiter's atmosphere, the Great Red Spot (GRS), returned by the *Voyager 2* probe in July 1979. Astronomers have observed the GRS for at least 300 years. It stays more or less in the same position in the atmosphere, just south of the equator in what is called the South Equatorial Belt. It varies in size over time, and at present it measures about 25,000 miles (40,000 km) across.

No one knew exactly what the GRS was until the *Voyager* probes visited the planet in 1979. They found that the GRS was a vast hurricane, with

gas spiraling anticlockwise around the center once every six days. The GRS towers about 5 miles (8 km) above the surrounding cloud tops. The intense red color is thought to be due to the presence of phosphorus compounds.

One or more white ovals are often seen in the vicinity of the GRS. They are similar, but smaller, hurricane-type systems, but are much less persistent. They last only a matter of months before others replace them.

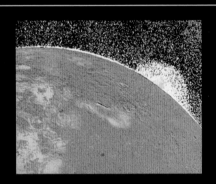

Volcanic Moon

A volcano erupts on the limb of Jupiter's moon Io. Sulfur fumes and dust shoot up to a height of more than 100 miles (160 km). And molten sulfur from Io's interior pours out over the surface, making Io one of the most colorful of all moons. Io measures about 2,260 miles (3,640 km) across. It is one of the four Galilean moons of Jupiter, named for Galileo, who first spotted them. The others are Europa, Callisto, and Ganymede. With a diameter of 3,275 miles (5,270 km), Ganymede is the biggest **Moon in the whole Solar System.**

Seeing Red
Close-up of Jupiter's most outstanding feature, the Great Red Spot. It is a region of intense turbulence and storms, as clouds eddy this way and that around this super-hurricane.

Diameter at equator	Av. distance from Sun	Spins around in
88,850 miles (143,000 km)	483,000,000 miles (778,000,000 km)	9 h 55 m
	Circles Sun in 11.9 years	**Number of moons** 60+

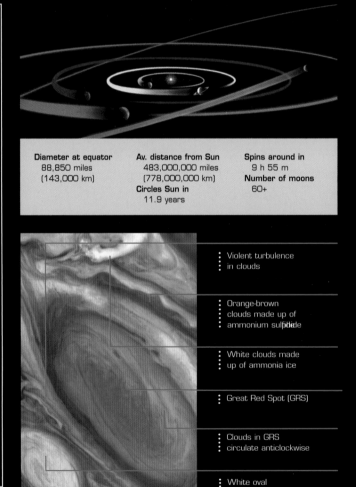

Violent turbulence in clouds

Orange-brown clouds made up of ammonium sulfide

White clouds made up of ammonia ice

Great Red Spot (GRS)

Clouds in GRS circulate anticlockwise

White oval

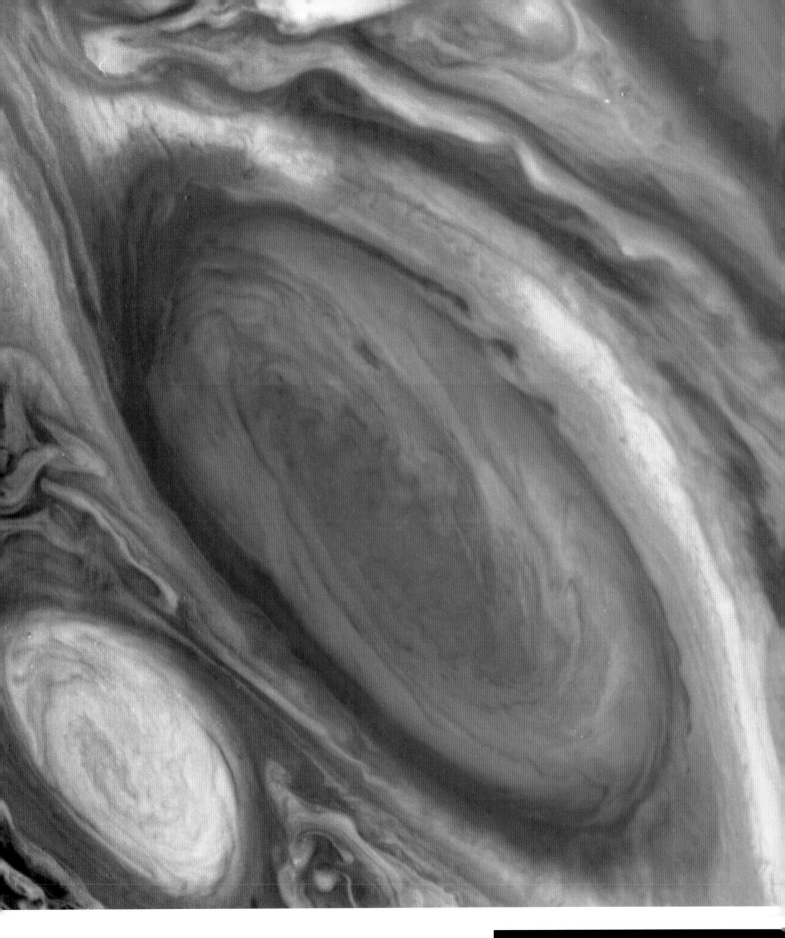

Saturn

Second in size only to Jupiter among the planets, Saturn is nine times bigger across than Earth. It is a jewel of a planet, surrounded by a beautiful shining ring system. The other gas giants—Jupiter, Uranus, and Neptune—also have rings, but they are but a pale imitation of Saturn's and are too faint to be visible from Earth.

Saturn is also similar to Jupiter in composition, being made up mainly of hydrogen and helium in the gaseous and liquid states. And it has similar cloud bands in its atmosphere like Jupiter, but much fainter. But it is the glorious ring system that dominates the planet.

Saturn tilts at an angle as it orbits the Sun about every 30 years. Over this period we see the rings at different angles. Sometimes we see them from the side (as happened in 1995); at other times, we see them in their most wide-open position.

The rings last appeared wide open in the spring of 2003, when the Hubble Space Telescope sent back the images in the picture. They are images taken at different wavelengths to bring out extra detail.

Ultraviolet light picks out the smaller particles, or aerosols, in the atmosphere. At visible and infrared wavelengths, absorption by methane in the atmosphere blocks out all but the uppermost layers in the atmosphere. Such studies help researchers understand the structure and dynamics of the atmosphere.

The picture also shows clearly the three main rings around Saturn. The outer A ring is separated from the B ring by a gap called the Cassini Division. Inside the B ring is a fainter C ring, which is almost transparent. The *Voyager* space probes discovered several other rings both inside and outside the main ring system.

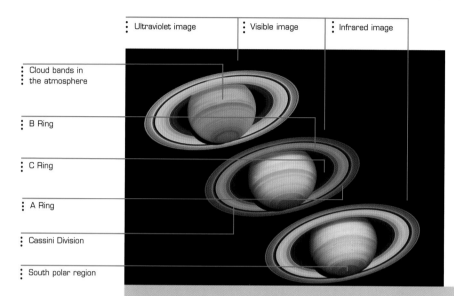

Ultraviolet image   Visible image   Infrared image

Cloud bands in the atmosphere

B Ring

C Ring

A Ring

Cassini Division

South polar region

Ring-a-ring

View of Saturn and its rings at three different wavelengths. The rings are made up of dusty water ice and range in size from tiny sand grains to boulders a yard (meter) or more across. Girdling Saturn's equator, they span a distance of some 170,000 miles (275,000 km), but in places they are only about 30 feet (10 m) thick.

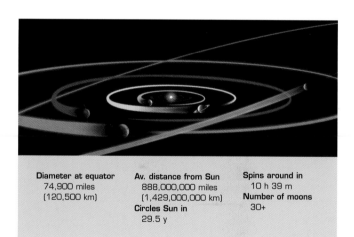

Diameter at equator
74,900 miles
(120,500 km)

Av. distance from Sun
888,000,000 miles
(1,429,000,000 km)
Circles Sun in
29.5 y

Spins around in
10 h 39 m
Number of moons
30+

Shepherd Moons

Saturn's 30+ moons vary widely in size. Biggest by far is Titan, which is more than 3,000 miles (5,000 km) across (see pages 152–153). Among the smallest are the so-called shepherd moons, found close to some of the rings. These moons are so called because they seem to keep ring particles in place just like shepherds herd their sheep. The picture shows the two shepherd moons Prometheus (right) and Pandora on either side of the narrow F ring.

Titan

Saturn's huge moon Titan was discovered in 1655 by the Dutch astronomer Christiaan Huygens. It is the second largest moon in the whole Solar System, after Jupiter's Ganymede. With a diameter of 3,200 miles (5,150 km), it is bigger than the planet Mercury.

But Titan has a much more important claim to fame than just size. It is the only moon in the Solar System that has an extensive atmosphere. The pressure of the atmosphere at the surface is more than one-and-a-half times the atmospheric pressure on Earth. The main gas in Titan's atmosphere is nitrogen. The other most important gas is methane, one of several hydrocarbons—hydrogen and carbon compounds—in the atmosphere.

Scientists estimate that a similar mix of gases was present in Earth's early atmosphere billions of years ago. They are hoping that detailed study of Titan's atmosphere could give clues as to how Earth's atmosphere—and even life on Earth—developed.

Titan was one of the main targets of the *Cassini-Huygens* probe, which went into orbit around Saturn on July 1, 2004. On October 26, the probe swooped within 730 miles (1,175 km) of the moon and returned the picture we see here.

The picture is a composite false-color image, made by combining images taken at ultraviolet and infrared wavelengths. Each wavelength reveals different kinds of details. Ultraviolet wavelengths, shown in blue, reveal the extent of Titan's atmosphere. Green colors represent infrared wavelengths and show where methane in the atmosphere absorbs light. Infrared also shows up denser methane clouds over the south pole.

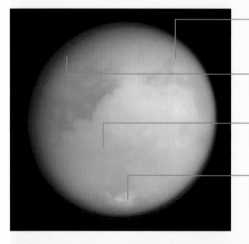

The northern hemisphere is much brighter than the southern

Titan's extensive atmosphere

Methane clouds gather over the south pole

Green showing where atmospheric methane absorbs light

Cassini's Titan
The Cassini orbiter snapped this full-face image of Titan in October 2004. It shows well the moon's atmosphere and the variation in brightness over the surface.

Surface Panorama

Titan was the specific target of the *Huygens* landing probe. *Huygens* cut loose from *Cassini* on Christmas Day 2004, and plunged into Titan's atmosphere on January 14, 2005. As it descended to Titan's surface, its imaging system snapped this 360-degree view of the surface from an altitude of about 5 miles (8 km). There seems to be a plateau region in the center of the image, and the white streaks visible could be ground fog. The probe drifted to a touchdown point in the dark area in the far right of the image, in a light wind blowing at a speed of around 4 mph (6 km/h).

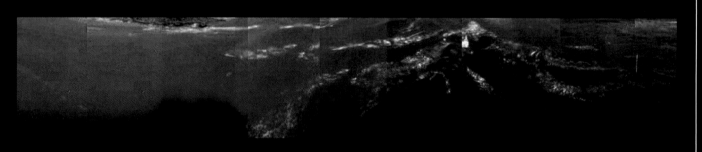

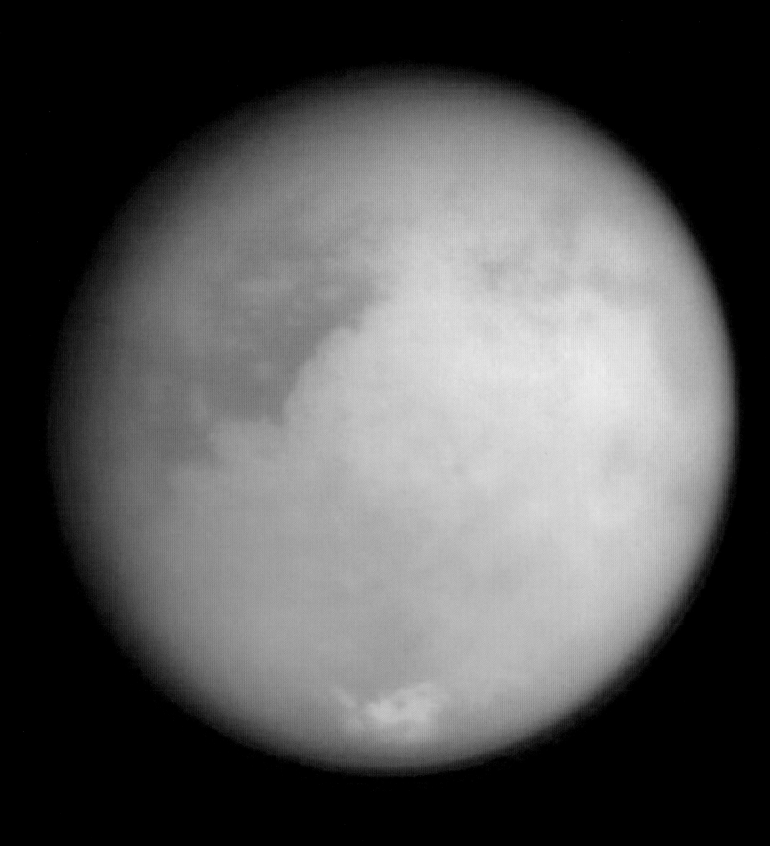

Uranus and Neptune

Uranus and Neptune are nowhere as big as Jupiter and Saturn, but they are still giant-sized compared with Earth. They are about four times bigger across than our own planet. Considering that they are separated from each other by a billion miles of space, they are remarkably similar—near twins in fact. They are almost the same size and have the same makeup.

Uranus has a bland greenish blue atmosphere, which in visible light shows virtually no features at all. The greenish blue color is the result of the absorption of other colors by methane in the atmosphere.

The Hubble Space Telescope has used infrared light to tease out some detail in the atmosphere (top picture). It has highlighted regions of haze and methane clouds. The picture also shows Uranus's ring system. The outermost and brightest of the 11 known rings is the Epsilon.

The *Voyager 2* probe provided a surprise when it targeted Neptune in August 1989 (bottom picture). Its cameras revealed that the planet is much more active than Uranus, even though it is so much farther away from the Sun.

The planet is a more intense blue than Uranus because it has more methane in its atmosphere. It shows bands of clouds, like Jupiter and Saturn, and it also has similar oval storm regions. The biggest one is the Great Dark Spot, named for its resemblance to Jupiter's Great Red Spot. But storm regions like this do not last as long as they do on Jupiter. When the Hubble Space Telescope scanned the planet a few years later, the spots had vanished.

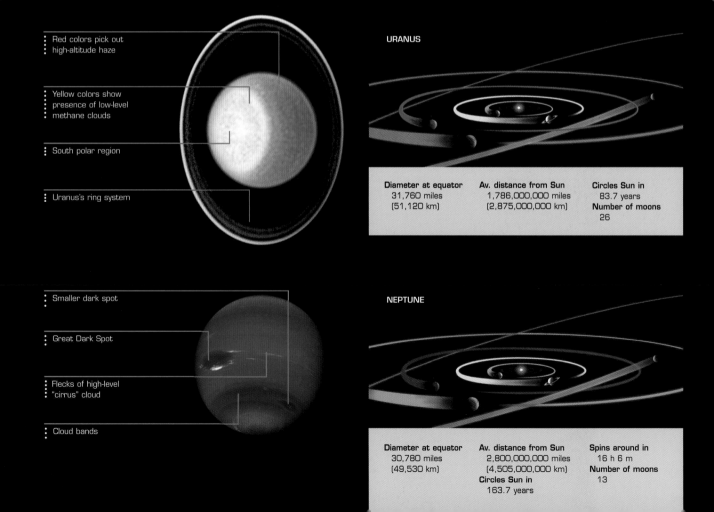

Red colors pick out high-altitude haze

Yellow colors show presence of low-level methane clouds

South polar region

Uranus's ring system

URANUS

Diameter at equator	Av. distance from Sun	Circles Sun in
31,760 miles (51,120 km)	1,786,000,000 miles (2,875,000,000 km)	83.7 years
		Number of moons 26

Smaller dark spot

Great Dark Spot

Flecks of high-level "cirrus" cloud

Cloud bands

NEPTUNE

Diameter at equator	Av. distance from Sun	Spins around in
30,780 miles (49,530 km)	2,800,000,000 miles (4,505,000,000 km)	16 h 6 m
	Circles Sun in 163.7 years	**Number of moons** 13

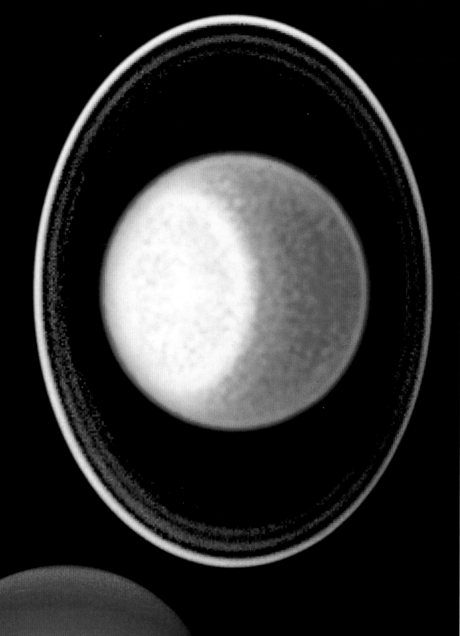

Hubble's Uranus

The Hubble Space Telescope returned this image of Uranus in infrared light. Infrared wavelengths pick out haze in the atmosphere and show well Uranus's ring system. The rings are made up of particles as black as soot.

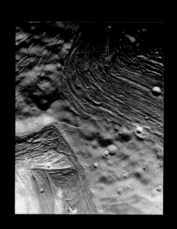

Remember Miranda

Only the five largest of the 26 or so moons of Uranus can be seen from Earth. The smallest of them, Miranda (diameter 290 miles/470 km), is by far the most interesting. It has the most incredible surface geology, as the picture shows. Rolling cratered plains suddenly give rise to peculiar grooved regions. This amazing geology might have come about if the moon had once been shattered in a collision with another body, and then reformed.

Voyager's **Neptune**

Dark cloud bands and vigorous oval storm regions characterized Neptune's atmosphere in 1989, when Voyager 2 flew past the planet. The probe swooped down to just 3,050 miles (4,900 km) above the cloud tops. It was a remarkable feat, given that Voyager had been traveling for more than 12 years into hitherto unexplored regions of space.

Triton and Pluto

Only about a fifth as big across as Earth, Pluto is a diminutive world, with the most eccentric, or elongated, orbit of any planet. It is a deep-frozen world, made up of rock and ice, and with a surface covered with frozen nitrogen and methane. It is quite unlike Uranus and Neptune, but seems to bear a close resemblance to Neptune's big moon Triton. The two bodies seem to belong to a class of bodies that astronomers call ice worlds.

The picture here shows a close-up view of Triton taken by *Voyager 2* when it flew past the moon in August 1989. It shows the amazing landscape of the south polar ice cap, which is probably made up of frozen nitrogen and other frozen gases. What is most remarkable is that the surface seems to be active. There appear to be active volcanoes or geysers, which are (or were recently) spewing out gaseous or liquid nitrogen, along with dark dust. Winds carry the dust across the surface, creating the dark streaks we see in the picture.

Triton and Pluto could be the largest, and closest, examples of a huge swarm of ice worlds that exist in the outer Solar System beyond Neptune. The region these ice worlds occupy is called the Kuiper Belt, and they are known as Kuiper Belt Objects, or KBOs.

More than 800 KBOs have already been discovered, mostly quite small. But, in 2002, a large one called Quaoar was found, orbiting some 4 billion miles (6 billion km) from Earth. In 2004, an even larger one was spotted, orbiting 8.5 billion miles (13.5 billion km) away. Named Sedna, it proved to have a diameter of about 1,000 miles (1,600 km). The media hailed Sedna as "the tenth planet," but astronomers preferred to call it a planetoid.

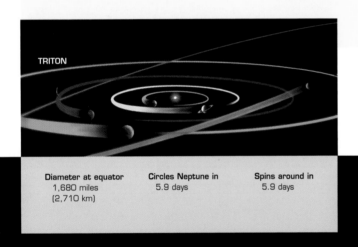

TRITON

Diameter at equator	Circles Neptune in	Spins around in
1,680 miles (2,710 km)	5.9 days	5.9 days

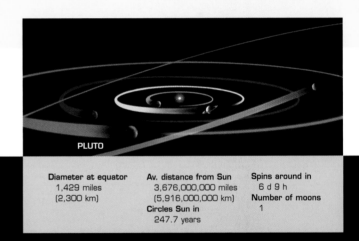

PLUTO

Diameter at equator	Av. distance from Sun	Spins around in
1,429 miles (2,300 km)	3,676,000,000 miles (5,916,000,000 km)	6 d 9 h
	Circles Sun in	Number of moons
	247.7 years	1

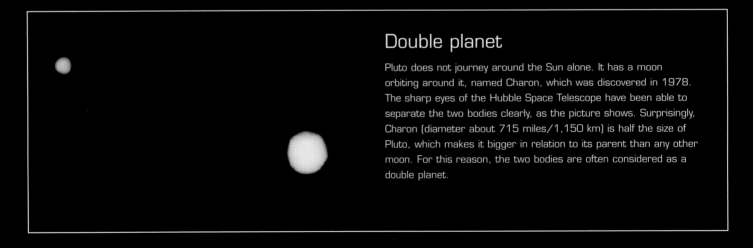

Double planet

Pluto does not journey around the Sun alone. It has a moon orbiting around it, named Charon, which was discovered in 1978. The sharp eyes of the Hubble Space Telescope have been able to separate the two bodies clearly, as the picture shows. Surprisingly, Charon (diameter about 715 miles/1,150 km) is half the size of Pluto, which makes it bigger in relation to its parent than any other moon. For this reason, the two bodies are often considered as a double planet.

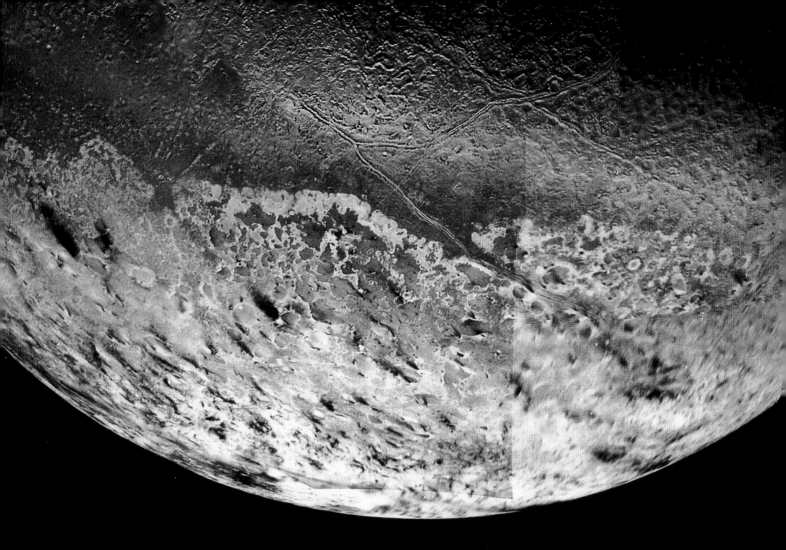

Triton's Icy Landscape

The Voyager 2 probe snapped this remarkable picture of Triton's south polar region. It shows an amazing landscape, with a pinkish snow cover and erupting geysers. The surface temperature is a frigid -390°F (-235°C).

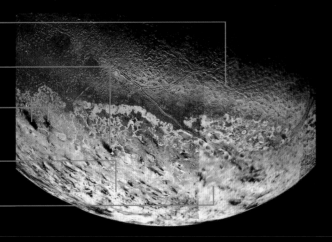

Wrinkled and ridged ice fields

Erupting "geyser"

Streaks of dark dust deposited on surface

South polar ice cap of Triton

Covering of pinkish snow

Quaoar

In 2002, astronomers spotted the biggest object found in the Solar System since Pluto was discovered in 1930. This is an artist's impression of what this ice world, named Quaoar, might be like. Quaoar is about half as big across as Pluto.

Comets

Comets are the mavericks of the Solar System. The brightest ones appear at irregular intervals and without warning from any part of the skies. At their most spectacular, they outshine the brightest stars and grow tails that can stretch for millions of miles against the background of stars.

But these apparently huge objects are insubstantial. What we see are tenuous clouds of gas and dust, illuminated by sunlight. The only substantial part of a comet is its nucleus, hidden in the bright head. This nucleus is generally only a few miles across. It is an icy lump made up mainly of dust and water ice—often described as a "dirty snowball."

Comets seem to originate in the outer depths of the Solar System in a spherical region known as the Oort Cloud. From time to time, one gets kicked out of the Cloud and starts traveling in toward the Sun. When it comes within Jupiter's orbit, the Sun's heat begins to evaporate some of the ice, releasing a cloud of gas and dust. Lit up by sunlight, this cloud becomes visible as a comet in our skies.

On July 23, 1993, U.S. astronomers Alan Hale in New Mexico and Thomas Bopp in Arizona were first to spot an unusually bright new comet well outside Jupiter's orbit. As convention dictates, the new comet was named Hale–Bopp. Less than four years later, Hale–Bopp was closing in on the Sun and had become one of the brightest comets of the century, clearly visible to the naked eye.

The author's photograph shows Comet Hale–Bopp at its brightest in March 1997. It shows up brilliantly against the dim orange sky of a long-gone sunset. It is located in quite a northerly position in the sky, as is evidenced by the close proximity of the constellation Cassiopeia. The comet has a bright head and a well-developed dust tail. (Its second tail—a bluish gas or ion tail—is not visible in this exposure.)

Target Jupiter

Strung out in line in space in May 1994 are the fragments of Comet Shoemaker-Levy 9, named for the astronomers who discovered it (Carolyn and Eugene Shoemaker and David Levy) the previous year. There are 21 fragments in all, stretched out over a distance of around 700,000 miles (1,100,000 km). They are rushing headlong toward Jupiter, and smashed into the planet in July 1994, creating gigantic fireballs that could be seen from Earth. No such collision had ever been witnessed before in the history of astronomy.

Backyard Comet

The author snapped this image of Comet Hale–Bopp from his back garden in the spring of 1997. Outshining almost all the stars, it hung in northwestern skies for many weeks and proved to be one of the brightest comets of the century.

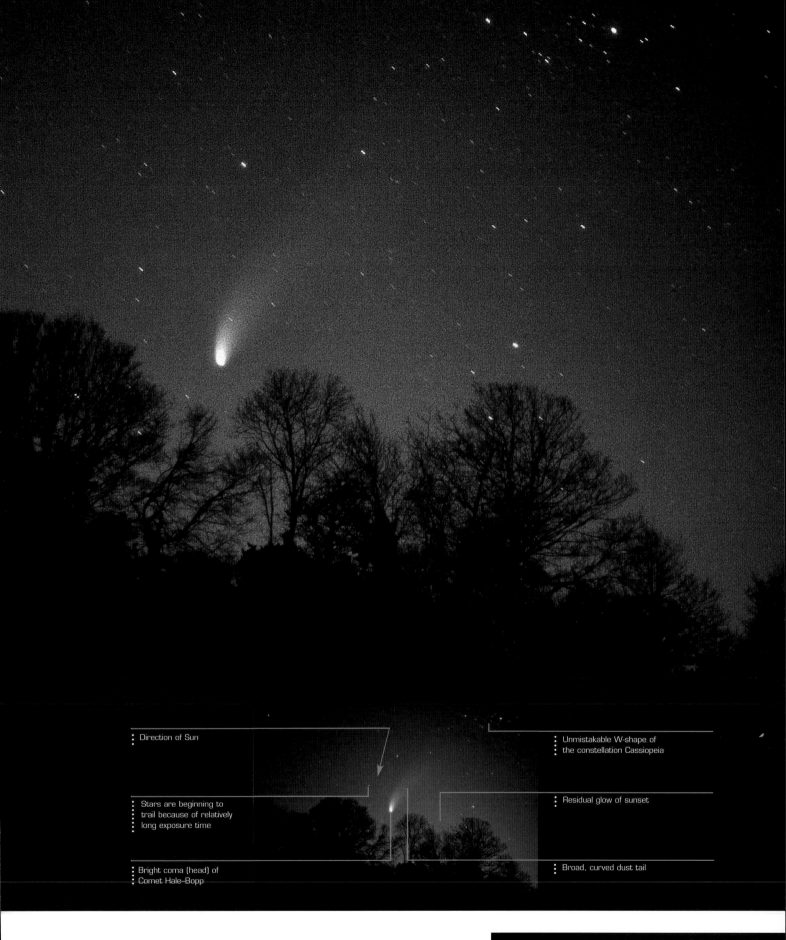

Direction of Sun

Unmistakable W-shape of
the constellation Cassiopeia

Stars are beginning to
trail because of relatively
long exposure time

Residual glow of sunset

Bright coma (head) of
Comet Hale–Bopp

Broad, curved dust tail

Asteroids

The asteroids, or minor planets, orbit in a band, or belt, in the huge expanse of space in the Solar System between the orbits of Mars and Jupiter. They are the remnants of primeval protoplanetary lumps that never managed to coalesce into a planet (see page 122).

Over time, the asteroids have smashed into one another, creating irregular lumps typically a few miles or tens of miles across. They seem to be made up of three main kinds of substance—rock, carbon, and metal. In some, rock predominates; in others, carbon or metal—iron and nickel in particular.

Most asteroids are too far away to be visible, and it was not until 1991 that close-up images of one were sent back by the *Galileo* probe, on its way to Jupiter. That asteroid was Gaspra. Two years later *Galileo* also imaged Ida. In 2000, the probe *Near-Shoemaker* went into orbit around Eros, and landed on it a year later.

The montage picture here shows images of Gaspra, Ida, and two hemispheres of Eros. Gaspra is quite a small asteroid, only around 12 miles (20 km) long. Like Ida and Eros, it appears to be one of the rocky asteroids. Ida is somewhat bigger—about 35 miles (55 km) across. Astonishingly, it has a miniature moon orbiting around it, named Dactyl, which is only about a mile (1.6 km) across.

But it was *Near-Shoemaker*'s mission to the 21-mile (33-km) long Eros that most excited the imagination. The probe did not just fly past the asteroid like *Galileo*, but also went into orbit around it and finally landed on it. This final and unplanned maneuver rates as one of the most incredible feats of space navigation ever.

The images of Eros illustrated show two hemispheres of the asteroid, snapped from a distance of around 220 miles (350 km). They show a strange saddlelike feature (Himeros) and also a comparatively large crater (Psyche).

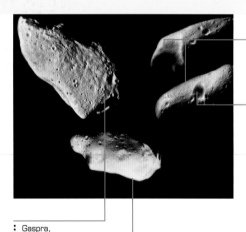

Two hemispheres of Eros, asteroid 433

Crater Psyche

Gaspra, asteroid 951

Ida, asteroid 243

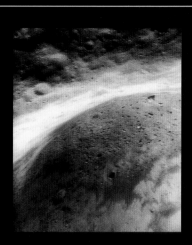

Eros in Close-up

Near-Shoemaker went into orbit around Eros on February 14 (Valentine's Day), 2000. (It was an appropriate day, because Eros was the son of the Greek goddess of love, Aphrodite.) While the probe was in orbit around Eros, it snapped this picture of the floor of the prominent crater Psyche. The relatively smooth floor is peppered with tiny craters made by the impact of debris falling from space. And it is covered with a kind of dusty soil. The rugged wall of the crater is visible at the top of the picture.

Shapeless Lumps

A montage of three of the few asteroids that have been photographed from space probes. They are irregular lumps of rock, pitted with a multitude of mainly tiny craters. Billions of such lumps are found within the asteroid belt.

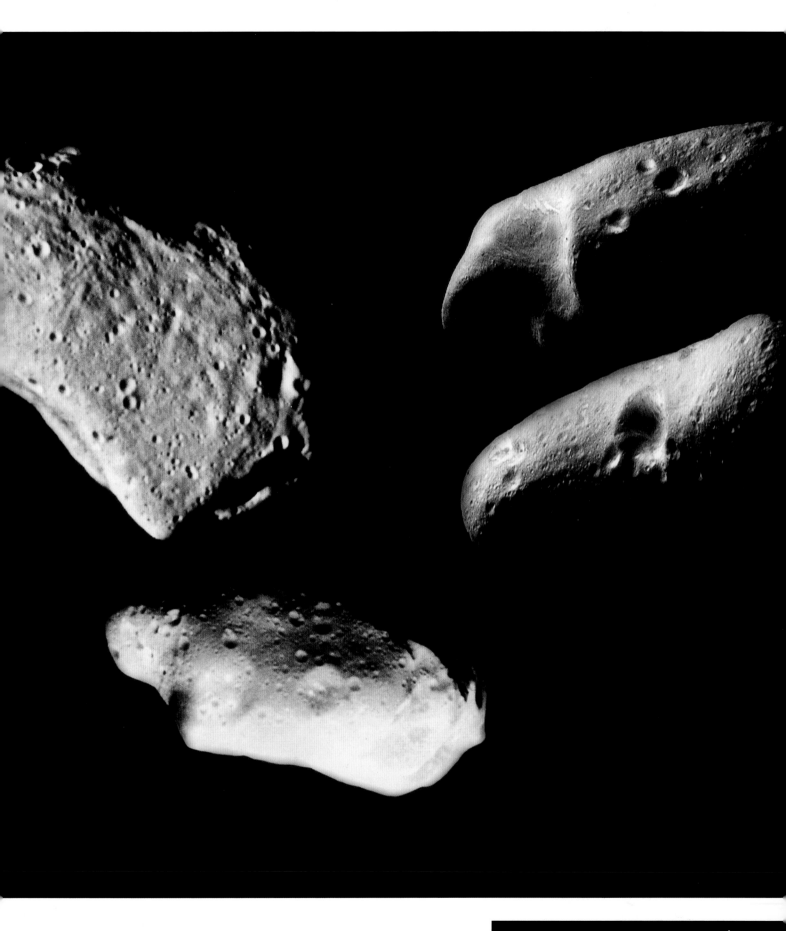

Mapping the Skies

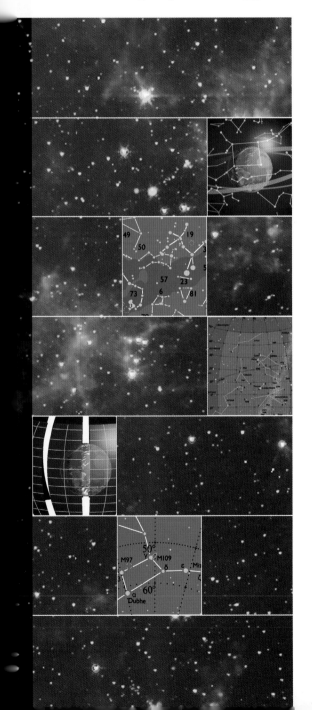

Ancient astronomers sought to comprehend the star-studded heavens thousands of years ago. They soon learned to recognize the constellations of bright stars and noticed how they appeared and disappeared from season to season. They envisaged the heavens as a great sphere surrounding Earth, a celestial sphere that they thought was the universe.

Today, we still use the constellations to help guide us across the heavens, and the geometry of the imaginary celestial sphere to prepare maps that pinpoint the stars in the night sky. This section presents the night sky in a series of monthly maps, which show how the constellations change throughout the year. Keyed into the maps are the locations of the images featured in Chapters 3 and 4.

Constellations and the Celestial Sphere

When we look up at the sky on a clear night, the heavens seem to form a huge dark dome above our heads. It is the same wherever we go on Earth. Earth seems to be set in the middle of a great spherical ball. The pinpoints of light that we call the stars seem to be stuck on the inside of this ball, which is otherwise impenetrably black.

Ancient astronomers believed in this great Earth-enveloping ball, which they named the "celestial sphere." Earth was fixed at the center of the celestial sphere; it didn't move, but the sphere did. The sphere rotated around Earth once a day, causing the stars to wheel across the heavens during the night from east to west.

We know now, of course, that there is no such thing as a celestial sphere. The stars all lie at different distances from us in the void of space, which seems to go on for ever and ever. We also know that it is not the starry heavens that rotate once a day but Earth, spinning on its axis from west to east.

Celestial Latitude and Longitude

Nevertheless, astronomers find the concept of a celestial sphere very useful, because they can use simple sphere geometry to pinpoint the positions of stars in the sky. They use a method similar to the one geographers use to locate places on Earth.

A place on Earth is located by two reference points—latitude and longitude. The latitude of a place is a measure (in degrees) of its distance from Earth's Equator. The longitude of a place is a measure (in degrees) of its distance from a particular north–south line (running through Greenwich, England).

Astronomers pinpoint the location of a star in a similar way, by its celestial latitude and celestial longitude. Celestial latitude, or declination (δ), is measured in degrees.

Celestial longitude, or right ascension (R.A.), is measured in hours of sidereal (star) time.

The Constellations

The bright stars that appear stuck to the inside of the celestial sphere form patterns we call the constellations. Ancient astronomers named the constellations after figures they thought the patterns resembled (see page 30).

In astronomy today, "constellation" does not just mean a pattern of bright stars that might form a recognizable figure, but refers to a particular area of sky—of the celestial sphere—that includes the pattern. In total, astronomers recognize 88 constellations.

Northern and Southern Constellations

For convenience, we can divide up the constellations into two: the ones that appear in the skies north of the celestial equator, in the northern half, or hemisphere, of the celestial sphere; and the ones that appear in the skies south of the celestial equator, in the southern celestial hemisphere.

The circular maps opposite show the constellations of the northern and southern hemispheres, with outline constellation patterns. The Key gives the constellation names in Latin and their English equivalents.

Exactly which constellations observers will see at any time will depend on where on Earth they are observing. Observers in, say, New York will be able to see all the northern constellations at some time during the year and many of the southern ones. But they will not be able to see far southern constellations, such as Crux, the Southern Cross.

Likewise, observers in Sydney, Australia, will be able to see all of the southern constellations at some time and some of the northern ones. But they will never be able to see the Big Dipper in far northern skies.

THE CELESTIAL SPHERE

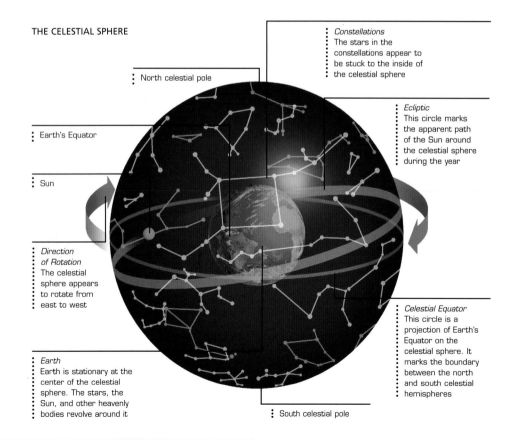

North celestial pole

Constellations
The stars in the constellations appear to be stuck to the inside of the celestial sphere

Ecliptic
This circle marks the apparent path of the Sun around the celestial sphere during the year

Earth's Equator

Sun

Direction of Rotation
The celestial sphere appears to rotate from east to west

Celestial Equator
This circle is a projection of Earth's Equator on the celestial sphere. It marks the boundary between the north and south celestial hemispheres

Earth
Earth is stationary at the center of the celestial sphere. The stars, the Sun, and other heavenly bodies revolve around it

South celestial pole

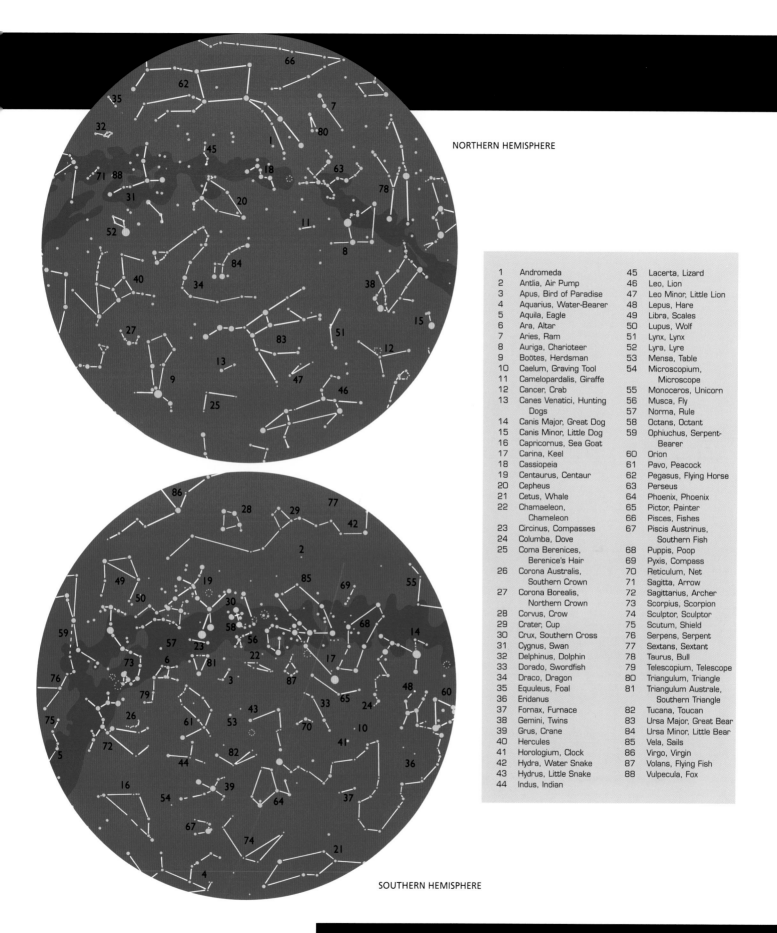

NORTHERN HEMISPHERE

1	Andromeda	45	Lacerta, Lizard
2	Antlia, Air Pump	46	Leo, Lion
3	Apus, Bird of Paradise	47	Leo Minor, Little Lion
4	Aquarius, Water-Bearer	48	Lepus, Hare
5	Aquila, Eagle	49	Libra, Scales
6	Ara, Altar	50	Lupus, Wolf
7	Aries, Ram	51	Lynx, Lynx
8	Auriga, Charioteer	52	Lyra, Lyre
9	Boötes, Herdsman	53	Mensa, Table
10	Caelum, Graving Tool	54	Microscopium,
11	Camelopardalis, Giraffe		Microscope
12	Cancer, Crab	55	Monoceros, Unicorn
13	Canes Venatici, Hunting	56	Musca, Fly
	Dogs	57	Norma, Rule
14	Canis Major, Great Dog	58	Octans, Octant
15	Canis Minor, Little Dog	59	Ophiuchus, Serpent-
16	Capricornus, Sea Goat		Bearer
17	Carina, Keel	60	Orion
18	Cassiopeia	61	Pavo, Peacock
19	Centaurus, Centaur	62	Pegasus, Flying Horse
20	Cepheus	63	Perseus
21	Cetus, Whale	64	Phoenix, Phoenix
22	Chamaeleon,	65	Pictor, Painter
	Chameleon	66	Pisces, Fishes
23	Circinus, Compasses	67	Piscis Austrinus,
24	Columba, Dove		Southern Fish
25	Coma Berenices,	68	Puppis, Poop
	Berenice's Hair	69	Pyxis, Compass
26	Corona Australis,	70	Reticulum, Net
	Southern Crown	71	Sagitta, Arrow
27	Corona Borealis,	72	Sagittarius, Archer
	Northern Crown	73	Scorpius, Scorpion
28	Corvus, Crow	74	Sculptor, Sculptor
29	Crater, Cup	75	Scutum, Shield
30	Crux, Southern Cross	76	Serpens, Serpent
31	Cygnus, Swan	77	Sextans, Sextant
32	Delphinus, Dolphin	78	Taurus, Bull
33	Dorado, Swordfish	79	Telescopium, Telescope
34	Draco, Dragon	80	Triangulum, Triangle
35	Equuleus, Foal	81	Triangulum Australe,
36	Eridanus		Southern Triangle
37	Fornax, Furnace	82	Tucana, Toucan
38	Gemini, Twins	83	Ursa Major, Great Bear
39	Grus, Crane	84	Ursa Minor, Little Bear
40	Hercules	85	Vela, Sails
41	Horologium, Clock	86	Virgo, Virgin
42	Hydra, Water Snake	87	Volans, Flying Fish
43	Hydrus, Little Snake	88	Vulpecula, Fox
44	Indus, Indian		

SOUTHERN HEMISPHERE

Mapping the Skies Month by Month

We do not see the same constellations in the skies all the time. As the months go by, some constellations disappear from view and others take their place. This happens because Earth travels in orbit around the Sun once a year.

Every month, as Earth travels a little farther in its orbit, we look out at night onto a slightly different part of the celestial sphere. After a year, when Earth has completed its orbit around the Sun, we look out at night onto the same part of the celestial sphere we saw 12 months before.

We can follow these changing aspects of the celestial sphere—the night sky—during the year in the series of star maps that follow.

For interest, in addition, we have keyed into the maps the location of the nebulae, clusters, galaxies, and other sky objects featured in the images presented in Chapters 3 and 4.

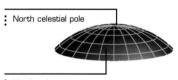

: North celestial pole

: North polar map

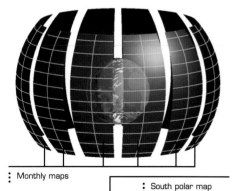

: Monthly maps

: South polar map

: South celestial pole

Mapping the Sky

To map the whole sky, we divide up the main body of the celestial sphere into 12 segments (see diagram, left). Each segment represents a part of the sky visible each month of the year. This provides us with 12 monthly maps—January through December.

To complete the picture, two further maps are needed, representing the northern and southern "caps" of the celestial sphere. They are centered, respectively, on the north and south celestial poles.

Mapping the Sphere

The diagram illustrates how we split up the celestial sphere for mapping purposes into 12 central segments and two polar "caps."

Around the Pole

Stars trail in circles around the north celestial pole in a long-exposure photograph. In the center is the Pole Star, Polaris, which scarcely trails at all.

Circumpolar Constellations

The north polar map (page 168) is centered on Polaris, the Pole Star, which lies close to the north celestial pole. For many observers in the northern hemisphere, the constellations close to the pole, such as Ursa Major and Cassiopeia, are always visible in the sky. We say they are circumpolar.

The south polar map (page 169) features stunning constellations such as Carina and Crux, the famous Southern Cross. They and the Large and Small Magellanic Clouds—our nearest galactic neighbors—are circumpolar for many southern observers.

The Monthly Star Maps

The map below is an example of the monthly star maps January (page 170) through December (page 181). It is actually the map for August. The annotations point out the kinds of details included in the maps, such as the symbols for sky objects like nebulae and clusters.

Each map shows the constellations visible near the meridian—the imaginary north–south line through the sky—at about 11:00 P.M. on about the 7th of each month.

Observers in Earth's northern hemisphere will see the constellation patterns this way when they look south at this time.

Observers in the southern hemisphere will need to turn the map the other way. This will then show how the constellations will appear when they look north at this time.

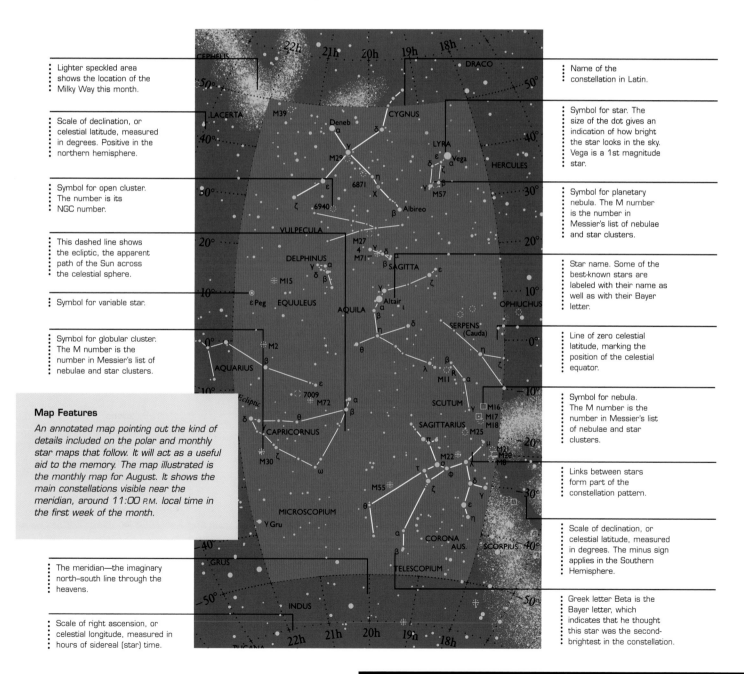

Lighter speckled area shows the location of the Milky Way this month.

Scale of declination, or celestial latitude, measured in degrees. Positive in the northern hemisphere.

Symbol for open cluster. The number is its NGC number.

This dashed line shows the ecliptic, the apparent path of the Sun across the celestial sphere.

Symbol for variable star.

Symbol for globular cluster. The M number is the number in Messier's list of nebulae and star clusters.

Map Features

An annotated map pointing out the kind of details included on the polar and monthly star maps that follow. It will act as a useful aid to the memory. The map illustrated is the monthly map for August. It shows the main constellations visible near the meridian, around 11:00 P.M. local time in the first week of the month.

The meridian—the imaginary north–south line through the heavens.

Scale of right ascension, or celestial longitude, measured in hours of sidereal (star) time.

Name of the constellation in Latin.

Symbol for star. The size of the dot gives an indication of how bright the star looks in the sky. Vega is a 1st magnitude star.

Symbol for planetary nebula. The M number is the number in Messier's list of nebulae and star clusters.

Star name. Some of the best-known stars are labeled with their name as well as with their Bayer letter.

Line of zero celestial latitude, marking the position of the celestial equator.

Symbol for nebula. The M number is the number in Messier's list of nebulae and star clusters.

Links between stars form part of the constellation pattern.

Scale of declination, or celestial latitude, measured in degrees. The minus sign applies in the Southern Hemisphere.

Greek letter Beta is the Bayer letter, which indicates that he thought this star was the second-brightest in the constellation.

North Polar Map

Constellations surrounding the celestial North Pole. For many northern observers, these constellations will be circumpolar.

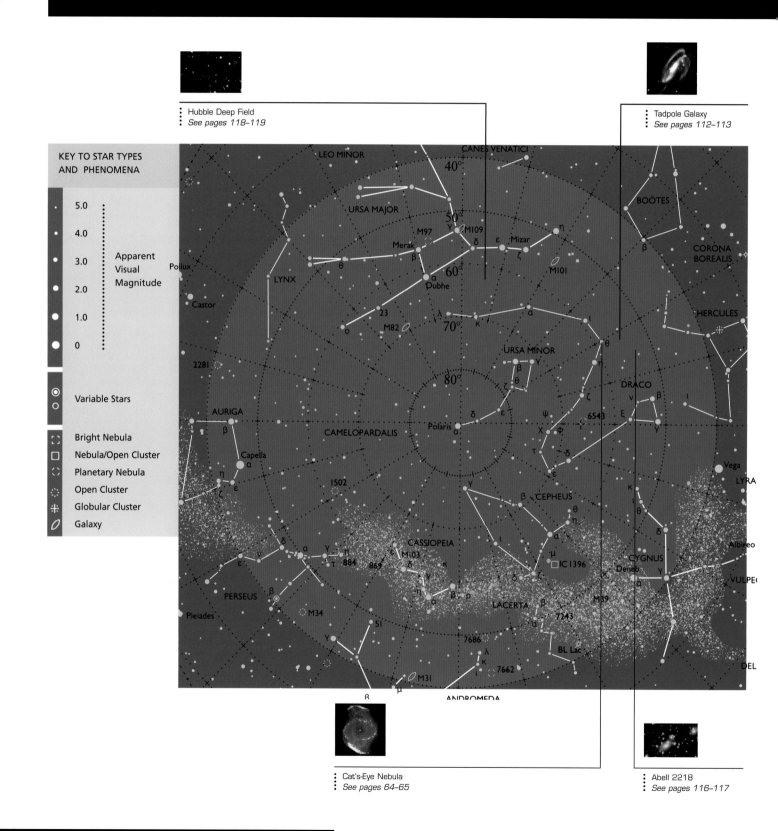

Hubble Deep Field
See pages 118–119

Tadpole Galaxy
See pages 112–113

KEY TO STAR TYPES AND PHENOMENA

5.0	
4.0	
3.0	Apparent Visual Magnitude
2.0	
1.0	
0	

◉ ○ Variable Stars

⬡ Bright Nebula
▢ Nebula/Open Cluster
◇ Planetary Nebula
⊙ Open Cluster
✳ Globular Cluster
⬭ Galaxy

LEO MINOR
CANES VENATICI
40°
BOÖTES
URSA MAJOR
50°
M97 γ M109
Merak δ ζ Mizar
β θ 60° M101
CORONA BOREALIS
Pollux
LYNX
Dubhe
HERCULES
Castor
23
70° κ λ α
M82
o
M103 ι
2281
URSA MINOR
β γ
ζ θ
DRACO
AURIGA
80° δ
ν β ι
β
ε Polaris α ζ
6543 ξ
Capella
α ψ φ
Vega
η ε τ δ
LYRA
ζ 1502
γ
κ
θ ι
β CEPHEUS θ δ
δ α γ η
δ γ η τ
884 μ IC 1396
Albireo
ε ν 869 M103 κ
CASSIOPEIA γ ι δ CYGNUS
ε δ η ζ β ρ Deneb γ
PERSEUS β α LACERTA β M39 VULPEC
M34 7243 α
Pleiades 51 7686 λ BL Lac DEL
γ κ 7662
M31
β μ
ANDROMEDA

Cat's-Eye Nebula
See pages 64–65

Abell 2218
See pages 116–117

South Polar Map

Constellations surrounding the celestial South Pole. For many
southern observers, these constellations will be circumpolar.

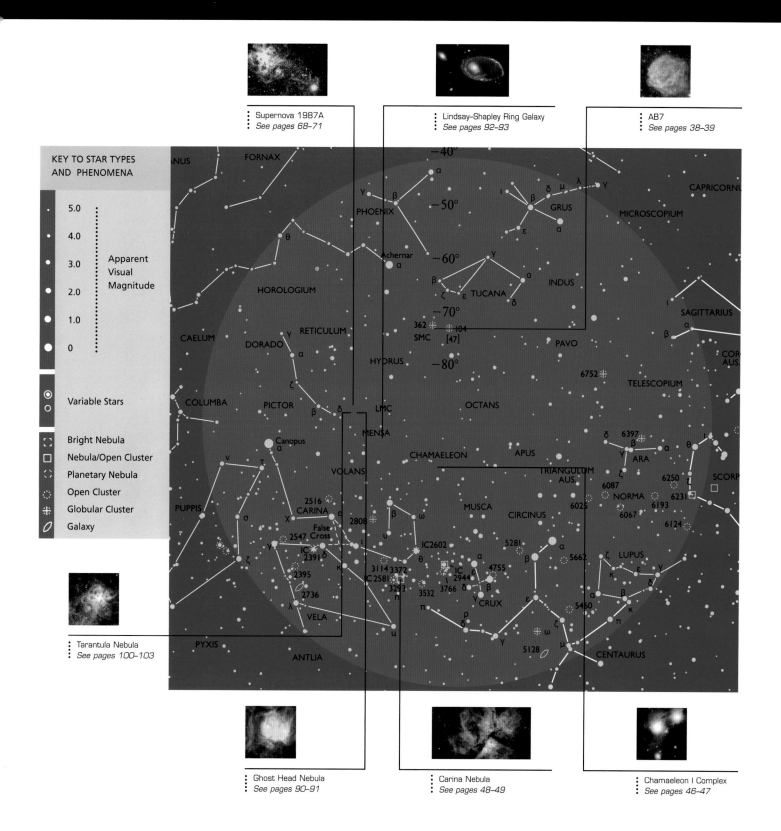

Supernova 1987A
See pages 68–71

Lindsay–Shapley Ring Galaxy
See pages 92–93

AB7
See pages 38–39

KEY TO STAR TYPES
AND PHENOMENA

5.0

4.0

3.0   Apparent
       Visual
2.0   Magnitude

1.0

0

Variable Stars

Bright Nebula
Nebula/Open Cluster
Planetary Nebula
Open Cluster
Globular Cluster
Galaxy

Tarantula Nebula
See pages 100–103

Ghost Head Nebula
See pages 90–91

Carina Nebula
See pages 48–49

Chamaeleon I Complex
See pages 46–47

Constellations visible near the meridian at around 11:00 P.M. local time during the first week of the month.

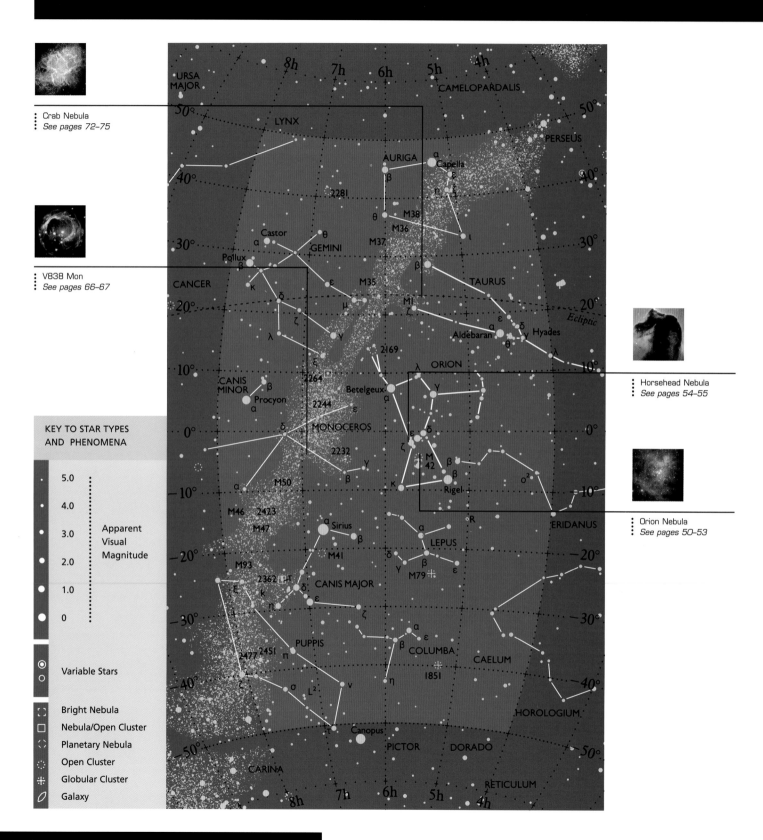

Crab Nebula
See pages 72–75

V838 Mon
See pages 66–67

Horsehead Nebula
See pages 54–55

Orion Nebula
See pages 50–53

KEY TO STAR TYPES AND PHENOMENA

- 5.0
- 4.0
- 3.0 — Apparent Visual Magnitude
- 2.0
- 1.0
- 0

Variable Stars

Bright Nebula
Nebula/Open Cluster
Planetary Nebula
Open Cluster
Globular Cluster
Galaxy

February Skies

Constellations visible near the meridian at around
11:00 P.M. local time during the first week of the month.

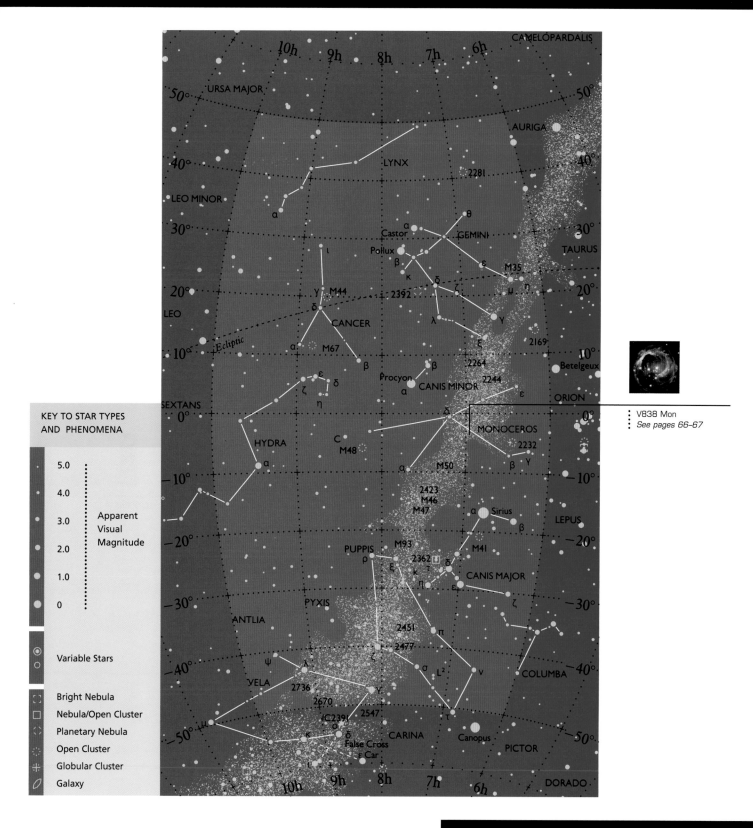

V838 Mon
See pages 66–67

KEY TO STAR TYPES AND PHENOMENA

5.0	
4.0	
3.0	Apparent Visual Magnitude
2.0	
1.0	
0	

Variable Stars

Bright Nebula
Nebula/Open Cluster
Planetary Nebula
Open Cluster
Globular Cluster
Galaxy

March Skies

Constellations visible near the meridian at around
11:00 P.M. local time during the first week of the month.

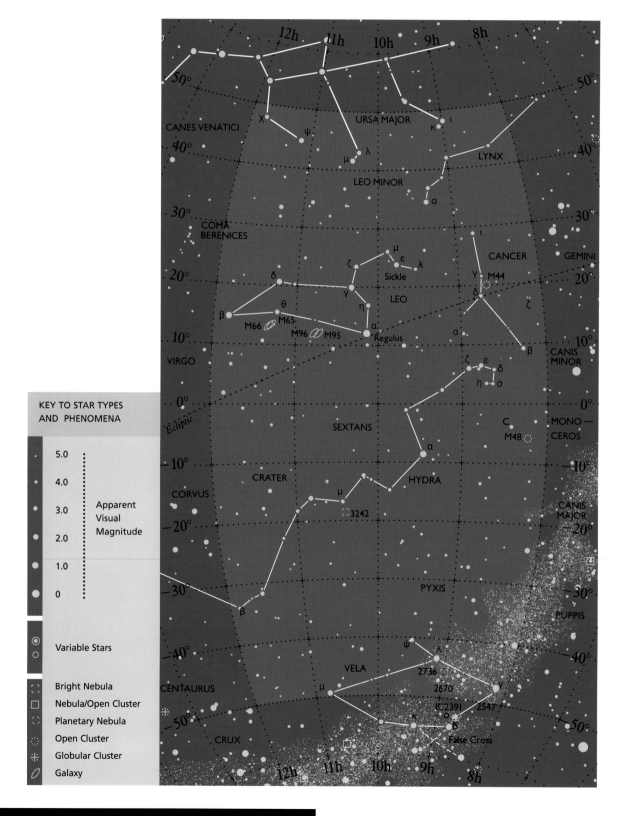

KEY TO STAR TYPES
AND PHENOMENA

Apparent
Visual
Magnitude

5.0
4.0
3.0
2.0
1.0
0

Variable Stars

Bright Nebula
Nebula/Open Cluster
Planetary Nebula
Open Cluster
Globular Cluster
Galaxy

April Skies

Constellations visible near the meridian at around
11:00 P.M. local time during the first week of the month.

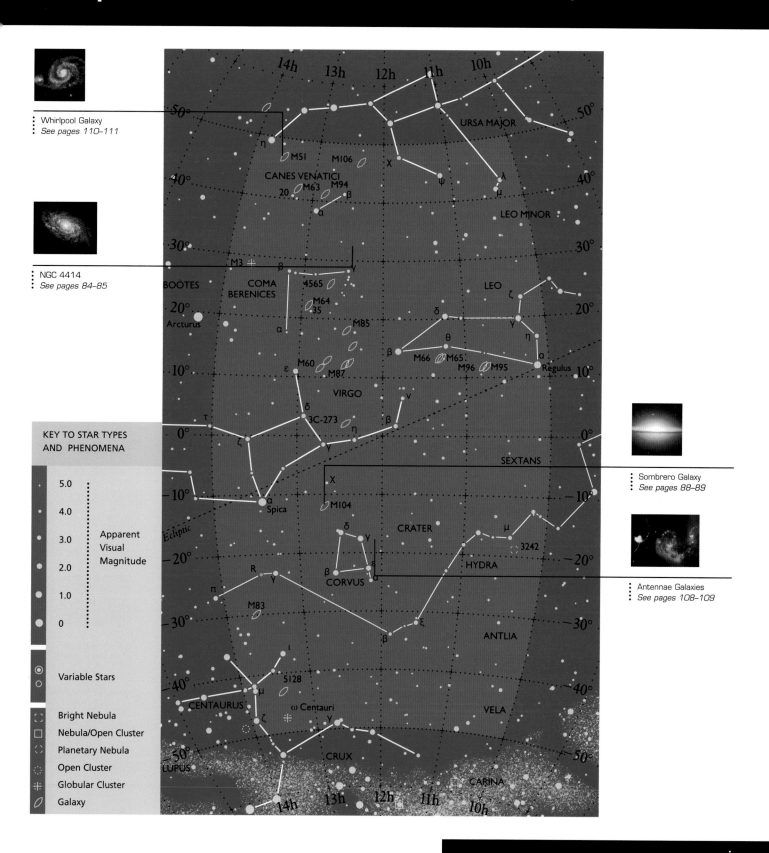

Whirlpool Galaxy
See pages 110–111

NGC 4414
See pages 84–85

Sombrero Galaxy
See pages 88–89

Antennae Galaxies
See pages 108–109

KEY TO STAR TYPES AND PHENOMENA

	Apparent Visual Magnitude
5.0	
4.0	
3.0	
2.0	
1.0	
0	

Variable Stars

Bright Nebula
Nebula/Open Cluster
Planetary Nebula
Open Cluster
Globular Cluster
Galaxy

14h 13h 12h 11h 10h

50° URSA MAJOR 50°

η M51 M106 χ
CANES VENATICI ψ
20 M63 M94 λ
β μ
α LEO MINOR
40° 40°

30° M3 β γ 30°

BOÖTES COMA 4565 LEO ζ
BERENICES δ η
20° Arcturus M64 γ 20°
35 θ α
α M85 β M66 M65 M96 M95 Regulus
ε M60 10° 10°
M87 ν
VIRGO
δ β
0° τ 3C-273 η 0°
ζ γ SEXTANS
X
-10° M104 -10°
α Spica δ γ
CRATER μ
β δ 3242
CORVUS α HYDRA
-20° R γ -20°
π
M83 ξ
-30° β ANTLIA -30°
ι
5128 VELA
-40° μ -40°
CENTAURUS ζ ω Centauri γ
-50° CRUX -50°
LUPUS CARINA
14h 13h 12h 11h 10h

May Skies

Constellations visible near the meridian at around
11:00 P.M. local time during the first week of the month.

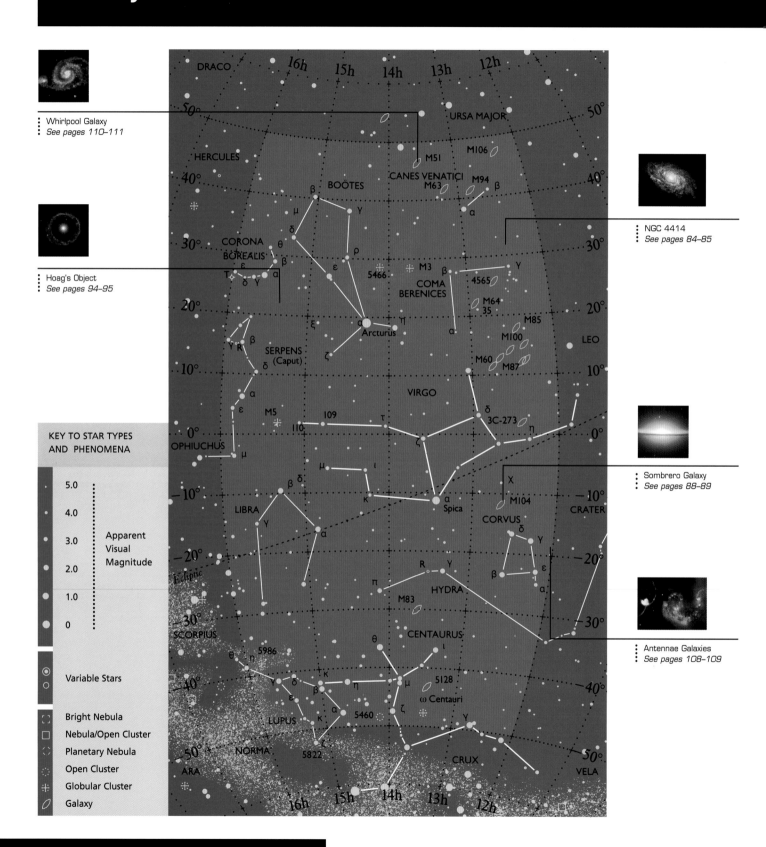

Whirlpool Galaxy
See pages 110–111

Hoag's Object
See pages 94–95

NGC 4414
See pages 84–85

Sombrero Galaxy
See pages 88–89

Antennae Galaxies
See pages 108–109

KEY TO STAR TYPES AND PHENOMENA

5.0	
4.0	
3.0	Apparent Visual Magnitude
2.0	
1.0	
0	

Variable Stars

Bright Nebula
Nebula/Open Cluster
Planetary Nebula
Open Cluster
Globular Cluster
Galaxy

DRACO · 16h · 15h · 14h · 13h · 12h
URSA MAJOR
HERCULES
M51 · M106
BOÖTES · CANES VENATICI
M63 · M94 · β
β · γ
μ
δ · ρ
θ · β · M3 · β · γ
CORONA BOREALIS · ε · 5466 · 4565
τ · δ · γ · α · COMA BERENICES · M64
ε · 35 · M85
ξ · η · M100
β · Arcturus · α · M60 · M87 · LEO
γ R · SERPENS (Caput) · ζ
δ · VIRGO
α · δ
ε · 3C-273
M5 · 109 · τ · η
110 · ζ · Spica
OPHIUCHUS · μ · μ · ι · χ
κ · M104 · CRATER
LIBRA · β δ · CORVUS · δ
γ · β · γ
α · ε
α
R · γ
π · HYDRA
Ecliptic · M83
SCORPIUS · CENTAURUS
θ
5986 · 5128
θ · η · κ · μ · ω Centauri
γ · δ · β · η · ζ
ε · α · 5460 · γ
LUPUS · κ · CRUX · VELA
ζ
NORMA · 5822
ARA
16h · 15h · 14h · 13h · 12h

June Skies

Constellations visible near the meridian at around 11:00 P.M. local time during the first week of the month.

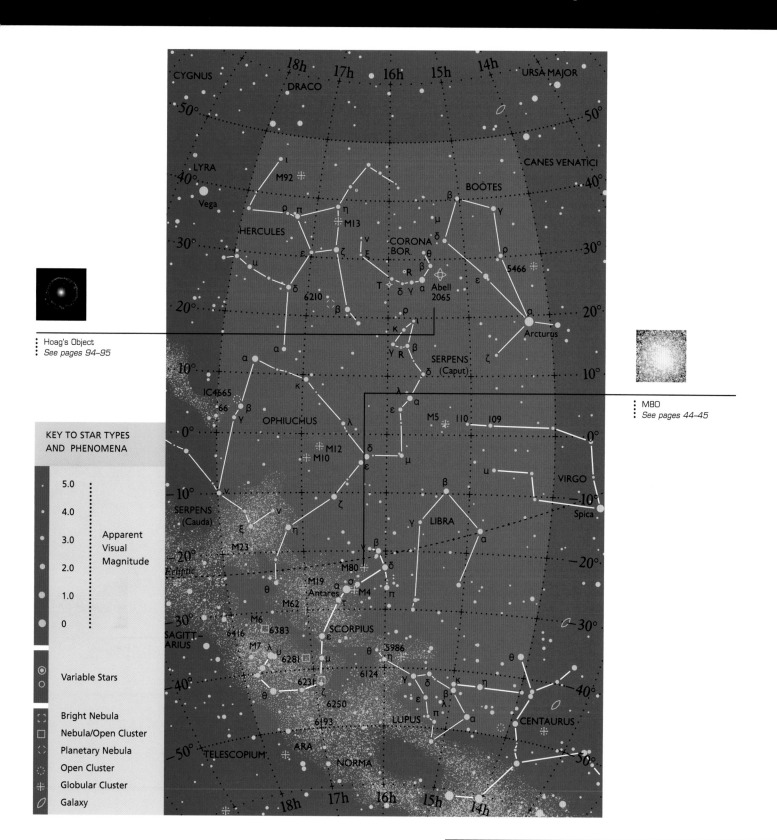

Hoag's Object
See pages 94–95

M80
See pages 44–45

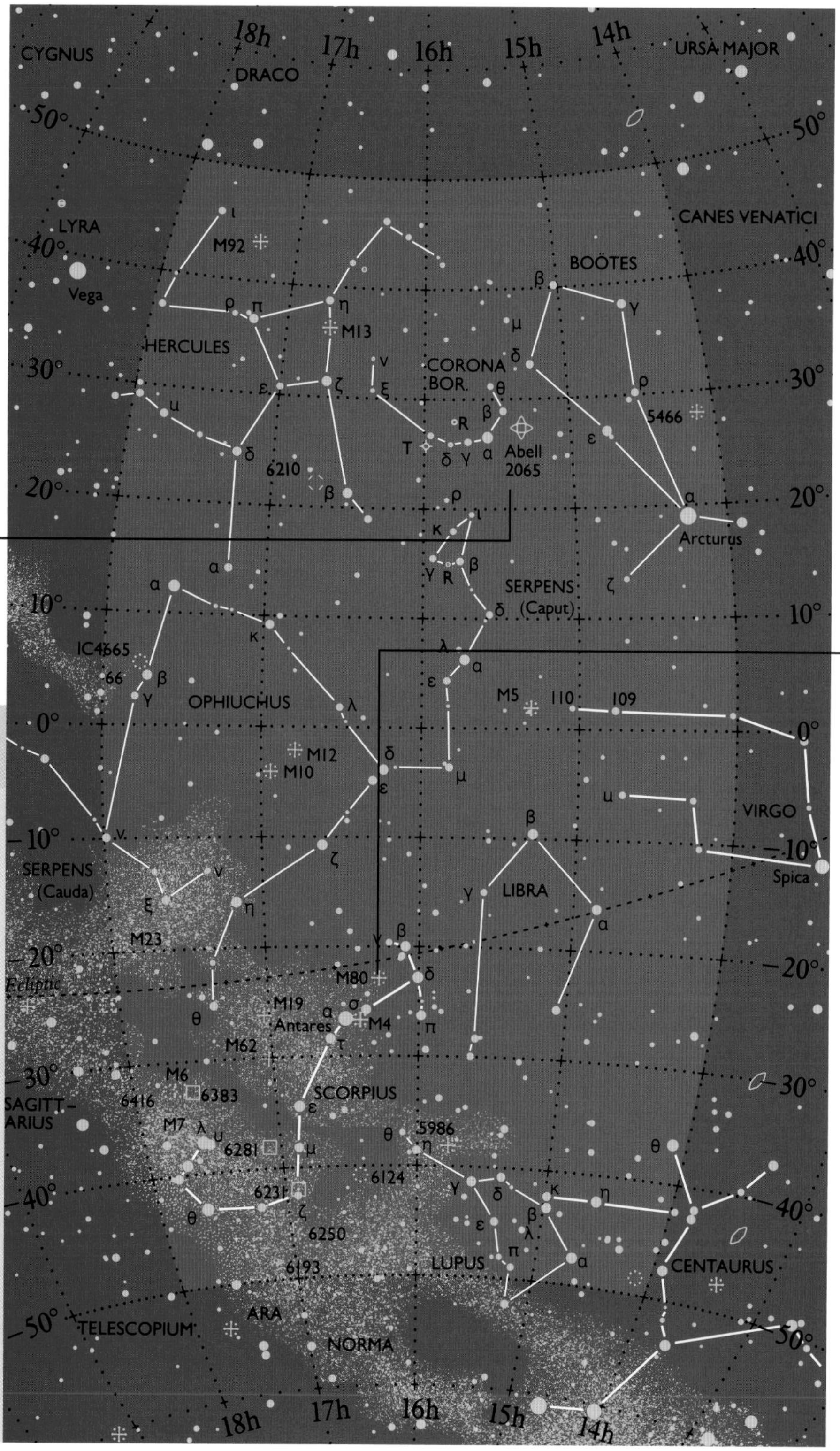

KEY TO STAR TYPES AND PHENOMENA

Apparent Visual Magnitude: 5.0, 4.0, 3.0, 2.0, 1.0, 0

Variable Stars

Bright Nebula

Nebula/Open Cluster

Planetary Nebula

Open Cluster

Globular Cluster

Galaxy

July Skies

Constellations visible near the meridian at around
11:00 P.M. local time during the first week of the month.

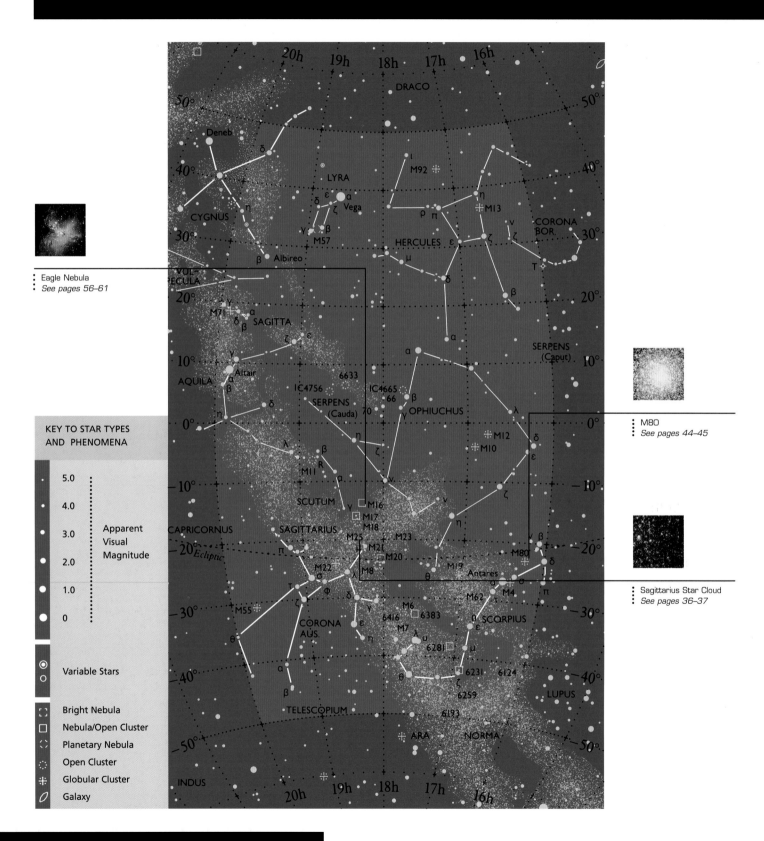

Eagle Nebula
See pages 56–61

M80
See pages 44–45

Sagittarius Star Cloud
See pages 36–37

KEY TO STAR TYPES AND PHENOMENA

•	5.0	
•	4.0	
•	3.0	Apparent
•	2.0	Visual
•	1.0	Magnitude
●	0	

◉ ○ Variable Stars

⬓ Bright Nebula
▢ Nebula/Open Cluster
⌣ Planetary Nebula
⊙ Open Cluster
✳ Globular Cluster
𝒪 Galaxy

August Skies

Constellations visible near the meridian at around
11:00 P.M. local time during the first week of the month.

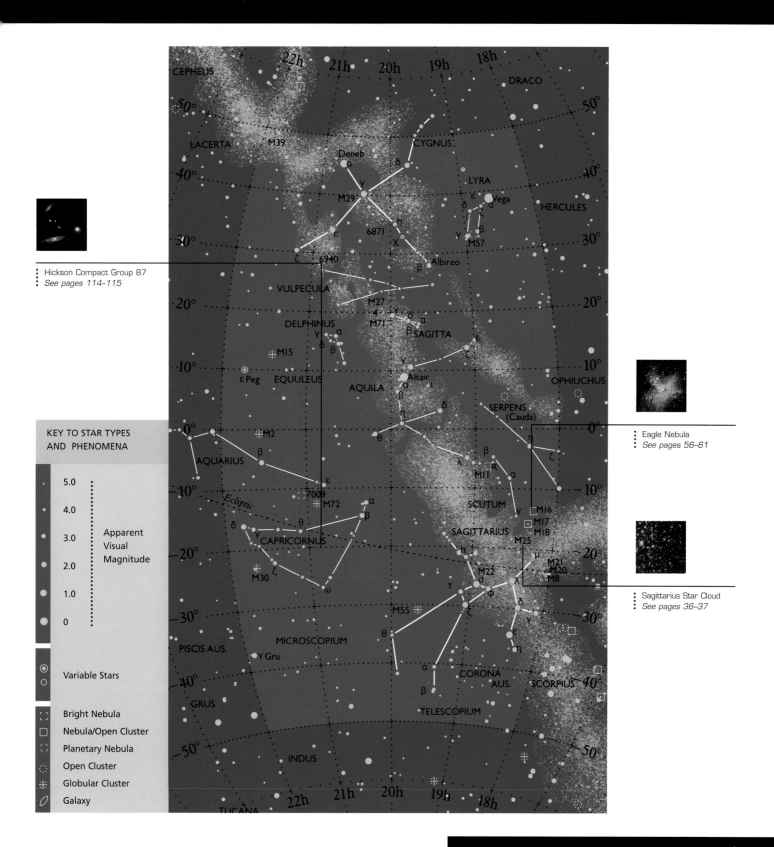

Hickson Compact Group 87
See pages 114–115

Eagle Nebula
See pages 56–61

Sagittarius Star Cloud
See pages 36–37

KEY TO STAR TYPES
AND PHENOMENA

5.0	
4.0	
3.0	Apparent
2.0	Visual
1.0	Magnitude
0	

⊙ ○ Variable Stars

⌞⌝ Bright Nebula
▢ Nebula/Open Cluster
◇ Planetary Nebula
⊙ Open Cluster
✣ Globular Cluster
⬭ Galaxy

September Skies

Constellations visible near the meridian at around
11:00 P.M. local time during the first week of the month.

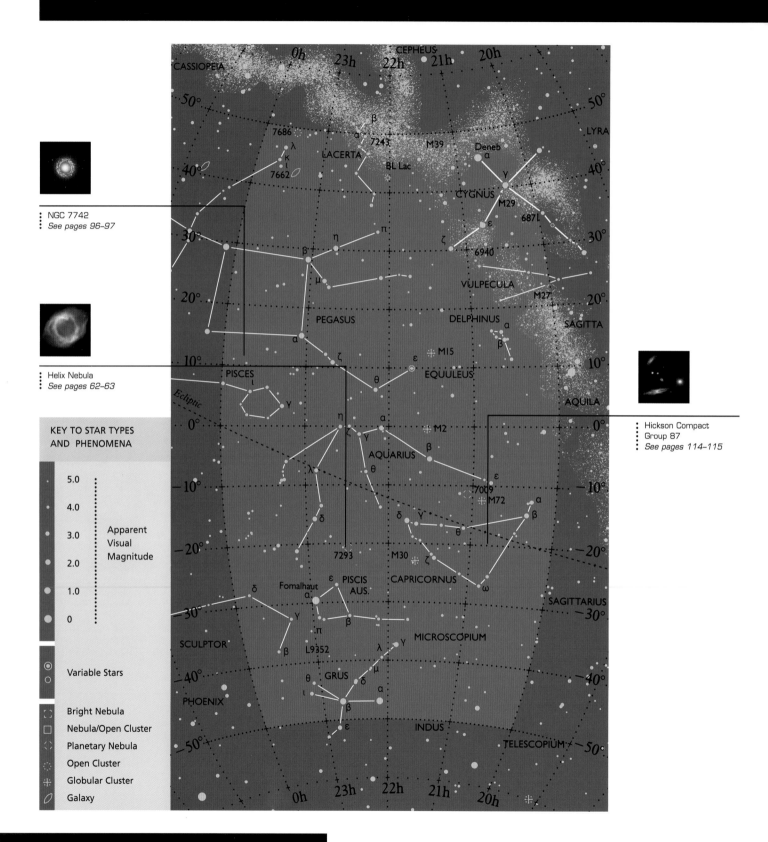

NGC 7742
See pages 96–97

Helix Nebula
See pages 62–63

Hickson Compact
Group 87
See pages 114–115

**KEY TO STAR TYPES
AND PHENOMENA**

•	5.0	
•	4.0	
•	3.0	Apparent Visual Magnitude
•	2.0	
●	1.0	
●	0	

◉ ○ Variable Stars

▫ Bright Nebula

▫ Nebula/Open Cluster

◇ Planetary Nebula

⬚ Open Cluster

✳ Globular Cluster

⬭ Galaxy

October Skies

Constellations visible near the meridian at around 11 P.M. local time during the first week of the month.

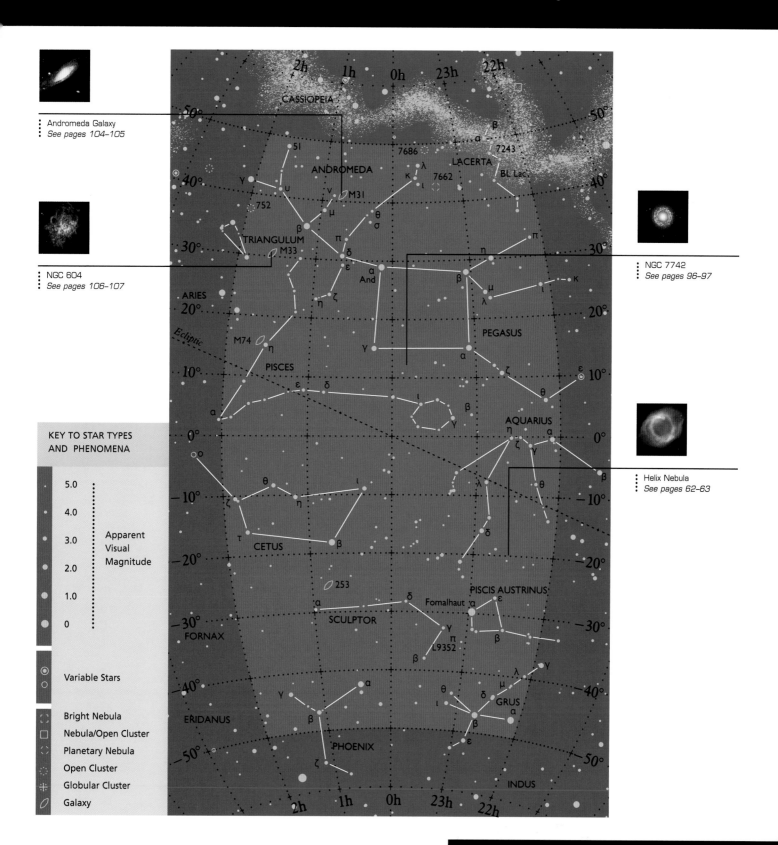

Andromeda Galaxy
See pages 104–105

NGC 604
See pages 106–107

NGC 7742
See pages 96–97

Helix Nebula
See pages 62–63

KEY TO STAR TYPES AND PHENOMENA

5.0	
4.0	
3.0	Apparent
2.0	Visual Magnitude
1.0	
0	

Variable Stars

Bright Nebula
Nebula/Open Cluster
Planetary Nebula
Open Cluster
Globular Cluster
Galaxy

November Skies

Constellations visible near the meridian at around 11:00 P.M. local time during the first week of the month.

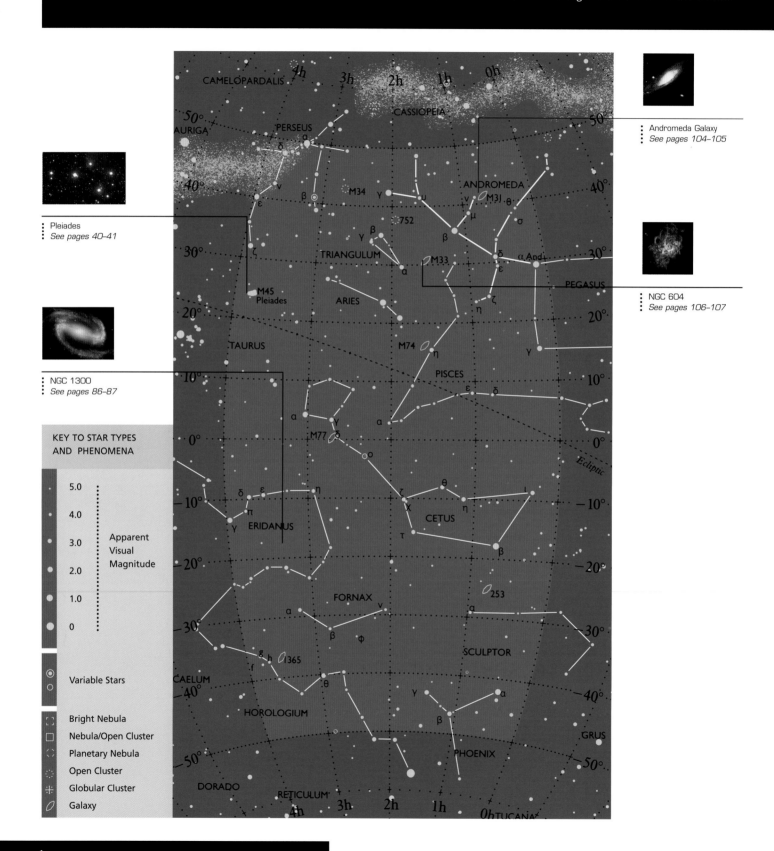

Andromeda Galaxy
See pages 104–105

NGC 604
See pages 106–107

Pleiades
See pages 40–41

NGC 1300
See pages 86–87

KEY TO STAR TYPES AND PHENOMENA

5.0	
4.0	
3.0	Apparent Visual Magnitude
2.0	
1.0	
0	

Variable Stars

Bright Nebula
Nebula/Open Cluster
Planetary Nebula
Open Cluster
Globular Cluster
Galaxy

December Skies

Constellations visible near the meridian at around
11:00 P.M. local time during the first week of the month.

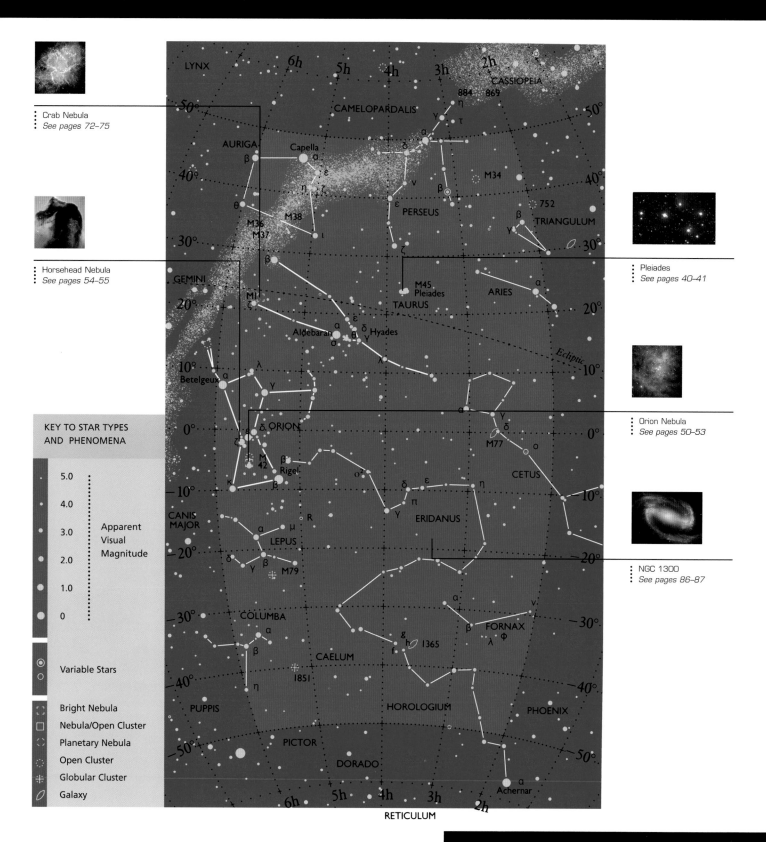

Crab Nebula
See pages 72–75

Horsehead Nebula
See pages 54–55

Pleiades
See pages 40–41

Orion Nebula
See pages 50–53

NGC 1300
See pages 86–87

KEY TO STAR TYPES AND PHENOMENA

Apparent Visual Magnitude

- 5.0
- 4.0
- 3.0
- 2.0
- 1.0
- 0

Variable Stars

Bright Nebula

Nebula/Open Cluster

Planetary Nebula

Open Cluster

Globular Cluster

Galaxy

Glossary

ABSOLUTE MAGNITUDE The true brightness of a star; the magnitude when viewed from a distance of 33 light-years.

ACCRETION DISK A disk of gas and dust that forms around newborn stars and black holes.

ACTIVE GALAXY A galaxy with an exceptionally high energy output, often as radio waves or X-rays, thought to originate from a massive black hole.

APPARENT MAGNITUDE The brightness of a star as we see it in the sky, measured on a scale from 1 (brightest) to 6 (faintest).

ASTEROID Also called minor planet; a chunk of rock or metal circling in space, mainly in a band (asteroid belt) roughly midway between the orbits of Mars and Jupiter.

ASTROLOGY An ancient belief that human lives and fortunes are somehow affected by the relative positions of the Sun and the planets among the stars.

ASTRONOMY The scientific study of the heavens and the heavenly bodies.

ATMOSPHERE The layer of gases that surrounds Earth and other planets. Also used as a unit of pressure—it is the pressure exerted by Earth's atmosphere at sea level.

ATOM The smallest unit of matter, having a central concentration of mass (nucleus) with electrons circling around it.

BARRED-SPIRAL GALAXY A kind of spiral galaxy that has a bar of stars through the nucleus.

BIG BANG The generally accepted theory of the origin of the universe, that it was created in a fantastic explosion about 13–15,000 million years ago.

BIG CRUNCH A theory that the universe will end in the opposite of a Big Bang, with all matter coming together and being crushed to nothing.

BILLION One thousand million— 1,000,000,000.

BINARY STAR A double-star system in which the two stars are bound by gravity and orbit around each other.

BLACK HOLE An area in space that has such powerful gravity that not even light can escape. It marks the death of a supermassive star or, in the center of galaxies, collapsed gas clouds.

BLAZAR An active galaxy in which we see the radiation coming straight from its core.

BLUE SHIFT A shift in the lines in the spectrum of a star toward the blue end, which indicates that the star is approaching us.

BROWN DWARF A small failed star, a body whose nuclear furnace has never lit up; it produces some heat but no light.

CCD Charge-coupled device, a silicon-chip device used in most modern telescopes to acquire images.

CELESTIAL SPHERE An imaginary dark sphere that appears to surround Earth, on the inside of which the stars appear to be fixed. The celestial sphere seems to revolve around Earth once a day, from east to west.

CEPHEID A variable star that changes brightness over a short period as regularly as clockwork; it can be used to estimate distances to far-off galaxies.

CHROMOSPHERE The colorful inner atmosphere of the Sun, which becomes visible from Earth during a total eclipse of the Sun.

CIRCUMPOLAR STARS For an observer, stars that are always visible in the night sky, circling around the celestial north pole or south pole.

CLUSTER A grouping of stars or galaxies. *See also* OPEN CLUSTER, GLOBULAR CLUSTER, SUPERCLUSTER.

COMA The bright head of a comet, which contains the nucleus.

COMET A lump of ice and dust that starts to shine when it approaches the Sun.

CONSTELLATION A group of stars that seem to be grouped together because they appear in the same part of the sky.

CORONA The outer atmosphere of the Sun, visible from Earth only during a total solar eclipse.

COSMOLOGY Study of the origin and evolution of the universe.

COSMOS An alternative name for the universe.

CRATER A pit in the surface of a planet or moon, usually made by the impact of a meteorite; some craters are volcanic in origin.

CRUST The solid outer surface of a planet or moon.

DARK MATTER Unseen matter that is thought to make up as much as 97 percent of the mass of the universe.

DARK NEBULA A cloud of interstellar matter that is not lit up by starlight. It can be detected when it blots out the light from stars behind it.

DOUBLE STAR A star that looks like a single star but is actually two stars close together. Some doubles are two stars that happen to appear together in our line of sight. Others are pairs of stars, called binary stars, that are actually close together.

ECLIPSE The passing of one heavenly body in front of another, which blots out its light. *See also* LUNAR ECLIPSE, SOLAR ECLIPSE.

ECLIPTIC The imaginary path of the Sun around the celestial sphere.

ELECTROMAGNETIC RADIATION Waves of energy given off, for example, by stars. They include gamma rays, X-rays, ultraviolet waves, visible light rays, infrared rays, microwaves, and radio waves. They all travel at the same speed—the speed of light—but differ in wavelength. Gamma rays are the shortest, radio waves the longest.

ELECTRON A tiny subatomic particle with little mass, present in all atoms; it has a negative electric charge.

ELEMENTS The basic "building blocks" of matter, which can't be broken down into simpler chemical substances.

ELLIPTICAL GALAXIES Galaxies of round or oval shape, which lack spiral arms.

ELLIPTICAL ORBIT An orbit that is oval in shape. The planets travel in elliptical orbits around the Sun.

EMISSION NEBULA A gas cloud that emits its own light; gas particles are triggered into emitting light by the intense radiation from nearby stars.

ENCOUNTER A meeting in space between a spacecraft and a planet or other heavenly body.

EVENING STAR The planet Venus (or Mercury) seen in the western sky after sunset.

EXPANDING UNIVERSE The theory that the universe is expanding, as evidenced by the recession of the galaxies.

EXTRASOLAR PLANET A planet in orbit around another star.

FALLING STAR A common name for a meteor.

FLY-BY A space mission in which a probe flies past a planet without going into orbit or landing.

GALAXY A "star island" in space. Our galaxy is the Milky Way, often called just the Galaxy. *See also* BARRED-SPIRAL GALAXY, ELLIPTICAL GALAXY, IRREGULAR GALAXY, RING GALAXY, AND SPIRAL GALAXY.

GAS GIANTS The planets Jupiter, Saturn, Uranus, and Neptune, which are giant-sized compared with the Earthlike planets and made up mainly of gas and liquid gas.

GLOBULAR CLUSTER A huge, ball-shaped mass containing hundreds of thousands of stars grouped closely together.

GRAVITATIONAL LENSING The distortion of the light from a distant object when it passes through a strong gravitational field, like that of a galaxy or galaxy cluster. It usually results in multiple images of the distant object.

GRAVITY The force of attraction that exists between bodies. The more massive a body is, the stronger is its gravitational attraction. Gravity holds the universe together.

GREENHOUSE EFFECT A situation in which the atmosphere traps heat like a greenhouse and causes global warming. Venus suffers from a galloping greenhouse effect.

HEAVENS The night sky, hence the heavenly bodies, the objects we see in the night sky.

HELIUM The second most common and second lightest element in the universe, after hydrogen. It is formed in stars by the nuclear fusion of hydrogen.

HYDROGEN The most common and lightest element in the universe, also the simplest; its atoms each contain just one proton and one electron.

IC NUMBER A means of identifying a star cluster, galaxy, or nebula; the number of the object in the Index Catalogue, an extension of the New General Catalogue (NGC) of Nebulae and Star Clusters.

INFRARED Electromagnetic radiation slightly longer in wavelength than visible red light. It is heat radiation.

INTERPLANETARY Between planets.

INTERSTELLAR Between stars.

INTERSTELLAR MATTER Matter in the form of gas and dust that is scattered in the space between the stars. It is sometimes visible as clouds, or nebulae.

IRREGULAR GALAXY A galaxy with no particular shape.

KUIPER BELT A region of space beyond Neptune's orbit, which contains millions of icy, cometlike bodies, known as Kuiper Belt Objects (KBOs).

LIGHT Electromagnetic radiation at visible wavelengths, which our eyes can detect.

LIGHT-YEAR The distance light travels in a year, about 5.9 million million miles (9.5 million million km). It is used as a unit for measuring distances in space. *See also* PARSEC.

LOCAL GROUP The small cluster of galaxies to which our own Galaxy belongs. It also includes the Andromeda Galaxy and the Magellanic Clouds.

LUNAR Having to do with the Moon.

LUNAR ECLIPSE An eclipse of the Moon, when it enters Earth's shadow in space.

MACHO Massive compact halo object, thought to be a form of dark matter present in the halo (spherical region) around a galaxy.

MAGNETIC FIELD The region around a star, galaxy, or planet where its magnetism acts.

MAGNITUDE The brightness of a star. Apparent magnitude is the brightness as we see it in the night sky. This is different from its true, or absolute, magnitude because no account is taken of how far away the star is. *See also* ABSOLUTE MAGNITUDE, APPARENT MAGNITUDE.

MANTLE The rocky layer of a planet or moon between the crust and core.

MARE A "sea" on the Moon, plural maria. Seas are in reality vast dusty plains.

MASS The amount of matter in a body.

MATTER The "stuff" that the universe is made of, composed of atoms or subatomic particles.

MESSIER NUMBER A means of identifying a star cluster, galaxy, or nebula; the number in a list drawn up by Charles Messier around 1781.

METEOR The streak of light we see in the sky when a meteoroid burns up high in the atmosphere.

METEORITE A lump of rock or metal that falls to the Earth from outer space; the remains of a large meteoroid.

METEOROID A piece of rock or metal traveling through space.

MILKY WAY A pale band of stars seen in the night sky. It represents a "slice" through our Galaxy, also called the Milky Way.

MINOR PLANET Another name for an asteroid.

MOLECULAR CLOUD An interstellar cloud made up of molecules, such as hydrogen. Stars are born from giant molecular clouds.

MOON The common name for the natural satellite of a planet. The Moon is Earth's only natural satellite.

MORNING STAR The planet Venus (or Mercury) in the eastern sky before sunrise.

NEAR-EARTH OBJECT (NEO) An asteroid whose orbit takes it close to Earth, sometimes dangerously so.

NEBULA A cloud of gas and dust in the space between the stars. *See also* DARK NEBULA, EMISSION NEBULA.

NEUTRON A subatomic particle found in the nucleus of every atom except hydrogen. As its name implies, it is electrically neutral.

NEUTRON STAR A very dense star, made up of neutrons packed tightly together.

NGC NUMBER A means of identifying a star cluster, galaxy, or nebula; the number of the object in the New General Catalogue of Nebulae and Star Clusters, published in 1888.

NOVA A star that suddenly flares up and brightens hugely. When a very faint distant star does this, it seems as if a new star has appeared; nova means new.

NUCLEAR FUSION A nuclear reaction in which light atoms (like hydrogen) fuse (join) together, releasing enormous energy.

NUCLEAR REACTION A process that involves the nuclei (centers) of atoms.

NUCLEUS The center of an atom, or a galaxy, or the small solid part of a comet.

OBSERVATORY A place in which astronomers work and from which they make their observations.

OORT CLOUD A huge spherical cloud more or less marking the edge of the Solar System, which is a reservoir of comets, proposed by Dutch astronomer Jan Oort in 1950.

OPEN CLUSTER A loose grouping of up to a few hundred, usually young, stars.

ORBIT The path in space one body follows when it circles around another. Most orbits are elliptical (oval) rather than circular.

PARALLAX The shift in position of a nearby object against a distant background when viewed from different points. Astronomers can use parallax shift to determine the distance to a few hundred of the nearest stars.

PARSEC The unit for measuring distances in space used by professional astronomers. It equals about 3.3 light-years.

PERIOD The length of time in a repeating cycle, such as the period of a comet—the time it takes to orbit the Sun.

PHASES The apparent changes in the shape of the Moon or other heavenly body (such as Venus) due to more or less of its surface being lit up by the Sun at different times.

PHOTOSPHERE The bright visible surface of the Sun or other star.

PLANET One of the nine main bodies that circle in space around the Sun, including Earth. Other stars have planets too, known as extrasolar planets.

PLANETARY NEBULA A cloud of gas and dust given off by a dying star of similar mass to the Sun.

PROBE A spacecraft that travels deep into space to explore other heavenly bodies, such as planets and their moons, comets, and asteroids.

PROMINENCE A fountainlike eruption of incandescent gas on the Sun, which often follows the invisible loops of the Sun's magnetic field.

PROPLYD Short for protoplanetary disk; a disk of matter around a newborn star from which planets might eventually form.

PROTON A subatomic particle found in all atoms; it has a positive electric charge.

PROTOSTAR An early stage in the birth of a star before nuclear processes have begun.

PULSAR A rapidly rotating neutron star that flashes pulses of radiation toward us, like a celestial lighthouse.

QUASAR A kind of active galaxy that pours out energy from a small central region. It is called quasar (for quasistellar) because it appears to be a starlike body, but it is much more remote than the stars and is as bright as hundreds of galaxies.

RADAR The technique of bouncing radio waves off objects, used in astronomy to map the surface of a planet such as Venus.

RADIO ASTRONOMY The branch of astronomy in which astronomers study the radio waves reaching Earth from space.

RADIO WAVES Electromagnetic radiation with the longest wavelengths.

RED DWARF One of the smallest stars; it is cool and shines faintly red.

RED GIANT A large red star near the end of its life. The Sun will one day become a red giant, swelling up to between 20 and 30 times its present size.

RED SHIFT A shift of the dark lines in the spectrum of a star or galaxy toward the red end, indicating that the object is moving away from us.

REFLECTOR A reflecting telescope, which uses mirrors to gather and focus incoming light.

REFRACTOR A refracting telescope, which uses lenses to gather and focus incoming light.

RETROGRADE MOTION Motion in the opposite direction from normal. Some of Jupiter's moons have retrograde motion, circling the planet in the opposite direction from the others.

RING GALAXY A galaxy consisting of a ring of star-forming material around a nucleus.

SATELLITE A small body that travels in orbit around a larger one; a moon. It is also the usual term for an artificial, man-made satellite.

SEA A flat plain on the Moon, properly called mare (plural, maria).

SEYFERT GALAXY An active galaxy with an exceptionally bright center.

SHEPHERD MOON A moon that orbits close to a planet's rings and appears to keep the ring particles in place.

SHOOTING STAR A common name for a meteor.

SOLAR Having to do with the Sun.

SOLAR ECLIPSE An eclipse of the Sun. In a total solar eclipse, the Moon covers all of the Sun; in a partial eclipse, it covers only part.

SOLAR FLARE A powerful explosion near the surface of the Sun that injects high-speed particles into the solar wind.

SOLAR SYSTEM The Sun and the bodies that travel with it through space. The main bodies are the planets, their moons, and the asteroids.

SOLAR WIND A stream of electrified particles given off by the Sun.

SPECTRUM A band of rainbow colors produced when sunlight or starlight is split into its separate wavelengths.

SPEED OF LIGHT The speed at which light and other electromagnetic radiations travel—around 186,000 miles (300,000 km) per second. It is the fastest speed possible.

SPIRAL GALAXY A galaxy with curved arms coming out of a bulging center, or nucleus. *See also* BARRED-SPIRAL GALAXY.

STAR A huge globe of hot gases that gives off energy as light, heat, and other radiation.

STELLAR Having to do with the stars.

STELLAR WIND A stream of particles that flows out from the surface of a star, as the solar wind flows from the Sun.

SUBATOMIC PARTICLE A particle smaller than an atom, such as a proton, neutron, and electron.

SUNSPOT A dark patch on the Sun's surface that is cooler than its surroundings. Sunspots come and go according to an 11-year cycle, called the solar (or sunspot) cycle.

SUPERCLUSTER A grouping of many clusters of galaxies occupying vast regions of space.

SUPERGIANT The largest type of star, typically hundreds of times bigger across than the Sun.

SUPERNOVA A catastrophic explosion in which a supergiant star blows itself to pieces (Type II). Even more powerful is the supernova (Type I) in which a white dwarf explodes.

TERRESTRIAL Having to do with Earth.

TERRESTRIAL PLANETS Planets made up mainly of rock such as Earth. They are Mercury, Venus, and Mars.

TRILLION One million million—1,000,000,000,000.

ULTRAVIOLET Electromagnetic radiation with a wavelength just shorter than that of visible violet light.

UNIVERSE Space and everything in it—galaxies, stars, planets, and energy.

VAN ALLEN BELTS Doughnut-shaped areas of intense radiation around Earth. Other planets have similar areas.

VARIABLE STAR A star that varies in brightness, usually because of a process going on inside it.

VOIDS Vast regions of empty space between the superclusters of galaxies that make up the universe.

WAVELENGTH The distance between the peaks or troughs of a wave motion, such as a wave of electromagnetic radiation. Electromagnetic wavelengths vary from billionths of a meter (gamma rays) to several kilometers (radio waves).

WHITE DWARF A small, very dense star about the same size as Earth. It marks a late stage in the life of a star of similar mass to the Sun.

WIMP Weakly interactive massive particle. Dark matter is thought to be made up mainly of WIMPs.

X-RAYS Penetrating electromagnetic radiation with a wavelength longer than gamma rays but shorter than ultraviolet.

ZODIAC An imaginary band in the heavens through which the Sun, Moon, and planets appear to travel; hence, the constellations of the zodiac.

Milestones in Astronomy

1543 Nicolaus Copernicus puts forward the idea of a Solar System.

1609 Johannes Kepler publishes the first of his laws of planetary motion, that planets move in elliptical orbits around the Sun.

Galileo builds a telescope and makes the first telescopic observations of the heavens.

1666 Isaac Newton formulates his laws of gravity and investigates the visible spectrum.

1781 William Herschel discovers Uranus, the first new planet found since ancient times.

1838 Using the parallax principle, Friedrich Bessel makes the first accurate measurement of stellar distance to the star 61 Cygni.

1846 Johann Galle discovers an eighth planet, Neptune.

1917 Completion of the 100-inch (2.5-m) Hooker Telescope at Mount Wilson Observatory.

1923 Edwin Hubble discovers a cepheid in the Great Nebula in Andromeda and establishes that it is a separate star system, an external galaxy.

1930 Clyde Tombaugh discovers a ninth planet, diminutive Pluto.

1931 Karl Jansky detects radio waves coming from the heavens, leading to the foundation of radio astronomy.

1948 The 200-inch (5-m) Hale Telescope completed at Mount Palomar Observatory.

1955 The first giant, steerable radio telescope (250 feet/82 m in diameter) is completed at Jodrell Bank Observatory, near Manchester, England.

1957 Russia's satellite *Sputnik 1* pioneers the Space Age when it is launched on October 4. A month later, *Sputnik 2* carries the first space traveler, a dog named Laika.

1958 The United States launches its first satellite, *Explorer 1*, on January 31, which discovers the Van Allen radiation belts around Earth.

1959 In September, Russia's *Luna 2* probe crashes on the Moon; in October, *Luna 3* returns the first pictures of the lunar far side, which is always hidden from Earth.

1961 On April 12, Russian cosmonaut Yuri Gagarin becomes the first human being to orbit Earth, once, in a Vostok capsule.

1962 In July, the United States launches the first successful deep-space probe, *Mariner 2*, which flies within 22,000 miles (35,000 km) of Venus, and reports on conditions there.

1965 In July, *Mariner 4* transmits the first close-up photographs of the planet Mars.

1969 In July, *Apollo 11* astronauts Neil Armstrong and Edwin Aldrin become the first humans to land on the Moon.

1971 *Mariner 9* becomes the first successful Mars orbiter in November.

1973 *Pioneer 10* (launched March 1972) flies past Jupiter in December, becoming the first probe to reach the giant planet.

1974 *Mariner 10* flies via Venus (in February) to Mercury.

1975 *Veneras 9* and *10* land capsules on Venus in October; they take the first close-up pictures of the planet's surface.

1976 Two *Viking* probes reach Mars, *Viking 1* in June, *Viking 2* in August. Orbiters survey the planet from orbit; landers take close-up pictures, report on Martian weather, and test soil for signs of life.

1979 *Voyager 1* (launched September 1977) reports back from Jupiter in March; *Voyager 2* (launched August 1977), in July.

1980 *Voyager 1* flies past Saturn in November.

1981 U.S. space shuttle makes its debut on April 12, when orbiter *Columbia* thunders into space. It returns to space in November, the first spacecraft ever to go into orbit twice.

1986 *Voyager 2* encounters Uranus in January.

1989 *Voyager 2* makes its final planetary encounter, with Neptune in August.

1990 Hubble Space Telescope launched from shuttle orbiter *Discovery* on April 24.

Magellan begins mapping the surface of Venus with radar in September.

1991 *Galileo* sends back the first images of an asteroid (Gaspra) in October.

2004 In January, *Mars Express* begins returning the best pictures yet of Mars from orbit. In the same month, the rovers *Spirit* and *Opportunity* begin roaming the Martian surface.

Stardust encounters *Comet Wild 2*, picking up cometary dust.

Planetoid Sedna discovered, the largest object found in the Solar System since Pluto.

Cassini–Huygens enters orbit around Saturn in July; in December releases *Huygens* landing probe.

2005 In January, *Huygens* plunges into Titan's atmosphere and sets down on the surface, sending back images of the surrounding landscape.

2006 *Stardust* scheduled to return samples of comet dust to Earth in January.

2011 *Messenger* scheduled to go into orbit around Mercury and map the whole surface.

James Webb Space Telescope, replacement for the Hubble Space Telescope, scheduled to be launched around this time.

2014 European probe *Rosetta* scheduled to encounter Comet Churyumov–Gerasimanko and land a probe on the surface.

On the Web

There is a wealth of astronomical information available on the Internet. This includes a huge photographic archive of images captured by the leading ground-based observatories, space observatories in orbit around Earth, and space probes journeying to planets, moons, asteroids, comets, and other bodies in the Solar System.

Astronomy Picture of the Day
APOD provides a stimulating start to the day for the astronomer. A different picture is featured each day, together with a brief explanation written by a professional astronomer.
http://antwrp.gsfc.nasa.gov/apod/

Hubble Space Telescope
Many of the images appearing in this book have been returned from the incomparable Hubble Space Telescope. Browse news releases and images at the following web site. Here, news and image information can be accessed for each year the Hubble Telescope has been in operation. Information can also be accessed according to category (Nebula, Galaxy, etc.).
http://hubblesite.org/newscenter/newsdesk/archive/releases

Very Large Telescope
Many of the images in the book were also taken by the ground-based Very Large Telescope of the European Southern Observatory in Chile, South America. Start with the VLT Astronomical Images Index, and select which category of object you want (Stars, Nebulae, etc.).
http://www.eso.org/outreach/ut1fl/astroimages.html

Gemini Observatory
The twin Gemini telescopes in the Northern and Southern Hemispheres are also returning excellent astronomical images. Go to this web site for the latest image and previous ones.
http://www.gemini.edu/index.php?option=com_gallery

SOHO
The Solar and Heliospheric Observatory has been returning the most spectacular pictures of the Sun since its launch in 1995. Follow the changing state of the Sun on the following web site. Links take you to a gallery of previous images.
http://sohowww.nascom.nasa.gov/

Planetary Photojournal
NASA's Jet Propulsion Laboratory maintains a Planetary Photojournal web site that features photographs of the planets and other Solar System bodies taken by a variety of different sources, mainly space observatories. Click on to whichever planet you wish to explore from the Solar System graphic. Images include those from the most recent probes, such as Cassini (Saturn).
http://photojournal.jpl.nasa.gov/index.html

Mars Express
Some of the best pictures of Mars ever have been taken by Europe's Mars Express orbiter. See them at http//:www.esa.int/SPECIALS/Mars_Express/index.html

Mission Information
Information about all missions by astronomy satellites and planetary probes and associated images is also readily available from numerous sites on the Internet. A good starting point for all NASA deep-space missions is the following web site. It includes features on current missions, like the Chandra X-Ray Observatory and the Spitzer Space Telescope, as well as lists of all other NASA missions.
http://www.nasa.gov/missions/deepspace/index.html

Voyager
The above site also reminds us that the most successful planetary missions of all time, NASA's *Voyager 1* and *2* missions to the outer planets and beyond, are still on-going. On January 5 and January 21, 2005, respectively, *Voyager 2* and *Voyager 1* celebrated their 10,000th day in space. Both are still going strong and returning valuable scientific data. For the latest *Voyager* information, go to
http://voyager.jpl.nasa.gov/

Europe's Deep Space Missions
Europe, under the auspices of the European Space Agency (ESA), conducts joint missions with NASA (such as Cassini–Huygens), and also conducts its own missions (such as Mars Express). Information about ESA missions is available from the following web site. Clicking on to a particular mission will highlight mission details and guide you to images.
http://sci.esa.int/science-e/www/area/index.cfm?fareaid=1#

Using Search Engines
Another way to access astronomical information on the Internet is to use a search engine, such as Google. This will provide you with a number of web sites, which may be specific to the subject you are searching for or more general. For example, to find out about the Spitzer Space Telescope (an infrared instrument) and access its images, enter "Spitzer Space Telescope" in the Search Engine window. Google, for example, hits the jackpot with its first entry, which directs us to the Spitzer Space Telescope home page:
http://www.spitzer.caltech.edu/

Now click on "Spitzer Site," which provides the latest information and a selection of Spitzer images. Now click on "Images," and you will be presented with thumbnails of images, which you can click to view larger versions.

Search engines are often not so good for more general subjects, such as the Moon. For example, entering "Moon" in a Google search provides nearly 78 million "hits." Entering "Apollo Moon-landing images" initially provides us with ten hits, but nine of these relate to the conspiracy theory that the Moon landings were a great hoax!

Index

Credits

The author and publishers would like to thank Spacecharts Photo Library for picture research and for supplying many of the photographs in the book.

They are also indebted to the following establishments for providing invaluable information and illustrative material:

Anglo-Australian Observatory

AURA, Associated Universities for Research in Astronomy

CERN, European Center for Nuclear Research

European Southern Observatory

European Space Agency (ESA)

Gemini Observatory

Hubble Heritage Team

Jet Propulsion Laboratory

Kitt Peak National Observatory

La Silla Observatory

Max Planck Institute for Radio Astronomy

NASA, National Aeronautics and Space Administration

NRAO, National Radio Astronomy Observatories

Palomar Observatory

Roque de los Muchachos Observatory

Royal Astronomical Society

Royal Greenwich Observatory

Space Telescope Science Institute

Very Large Array

Very Large Telescope

Time Life Pictures/Getty Images: p10

Photographs on the following pages were taken by the author:
p18, p158, p159, p166

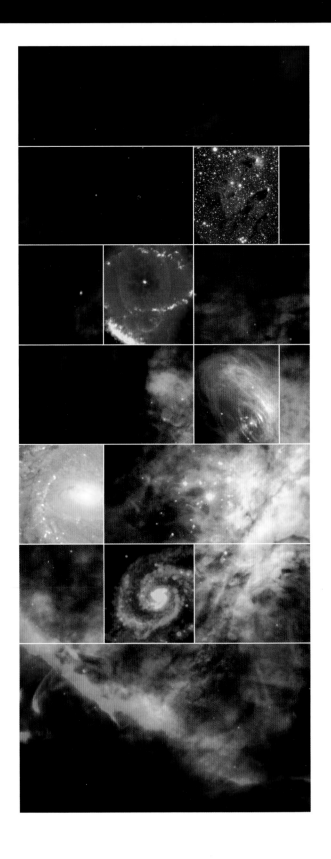